A Dictionary of Food Hygiene

R BOWMAN
and
E D EMMETT

STANLEY THORNES (PUBLISHERS) LTD

First published in 1992 by:
Stanley Thornes (Publishers) Ltd
Old Station Drive
Leckhampton
CHELTENHAM GL53 0DN
England

A catalogue record for this book is available from the British Library.

ISBN 07487 1405 7

Photoset in Sabon with Optima by Northern Phototypesetting Co. Ltd, Bolton
Printed and bound in Great Britain at The Bath Press, Avon

Contents

Entries, A–Z 1

Appendix 1: Food Law in Britain 204

Appendix 2: Scientific names 214

Appendix 3: Abbreviations 217

Appendix 4: Major causes of food poisoning and food-borne disease 219

Acknowledgements

Approved reference material for IEHO Food Hygiene courses.

Also endorsed by the Royal Institute of Public Health and Hygiene.

Rentokil Ltd.

Abattoir Large, often factory-type slaughterhouse, which may include a processing plant. The term is also commonly used to mean any slaughterhouse.

Aberdeen typhoid outbreak In 1964 an outbreak of typhoid fever occurred which involved hundreds of people in Aberdeen, Scotland. It was caused by canned corned beef produced in Argentina. The cans were cooled after processing with unchlorinated water from a river and it is suspected that the typhoid organisms entered the seams while they were expanded after heating. *See:* Canning, Infections (water-borne), *Salmonella typhi.*

Abrasion An area of grazed skin which is a potential site for bacterial growth leading to an infection. The covering of such areas with a suitable waterproof dressing is compulsory in food handling. *See*: Appendix 1: Food law, Personal hygiene, Pus.

Abrasives Various gritty substances used in the formulation of cleaning agents. An abrasive cleaner used incorrectly or on the wrong surface can cause scratches and surface damage which, although almost invisible to the naked eye, may harbour micro-organisms. This has been noted even with surfaces as durable as stainless steel.

Abscess A localised region of pus surrounded by inflamed tissue in an infected part of the body. It may be the cause of bacterial food poisoning if the pus is not contained. *See: Staphylococcus aureus*, Personal hygiene.

Accelerated Freeze Drying (AFD) A modern method of food dehydration.

- Small pieces of quick-frozen food are placed under vacuum and are then gently heated.
- Ice crystals in the food turn directly into water vapour without first melting. This effect is called sublimation.
- The food remains more or less the same size and after processing has a brittle and sponge-like character.
- During rehydration water enters faster and more completely than with conventionally dried products.
- The final quality is quite close to that of the fresh item.

Typical products include coffee, whole prawns, peas and other similar sized items. The main problems with the process are its expense and that the dried product is brittle and may powder if roughly handled. Also, the process will not destroy all bacteria and toxins. Shelf-life is long (months or more) as long as the packaging remains intact. The rehydrated food should be stored and treated as though it were fresh. *See*: Dehydration.

Acetic (ethanoic) acid A preservative used in the food industry. It inhibits organisms responsible for both food poisoning and food spoilage. It is more effective against yeasts and bacteria than moulds, and it is the active component in mayonnaise, pickles and many sauces. *See*: Acid, Food Preservation, pH.

Acetobacter A group of Gram-negative bacteria used in the manufacture of vinegar. They oxidise alcohol (ethanol) from, for example, wine or apple juice, producing acetic (ethanoic) acid. They can be responsible for spoiling beers and wines, producing a vinegary taste.

Achromobacter A group of Gram-negative bacteria found in the intestinal tract of man and animals. They can cause food spoilage of beer, dairy products, eggs, milk and other foods. (Many of the species of *Achromobacter* have been re-named as species of *Alcaligenes*.)

Acid A chemical substance which, when dissolved in water, releases hydrogen ions (H^+). Most bacteria cannot tolerate strong acid conditions. They grow best at around pH 6.5, but they can grow at pH values between 4 and 9. This sensitivity to strong acid conditions is used in the method of food preservation by pickling, where vinegar (a weak solution of acetic/ethanoic acid) is used to stop the growth of micro-organisms. In general, yeasts and moulds are more tolerant of acid conditions than bacteria, so foods with a low pH (below pH 4.5) are not usually spoiled by bacteria. They are more susceptible to spoilage by yeasts and moulds.

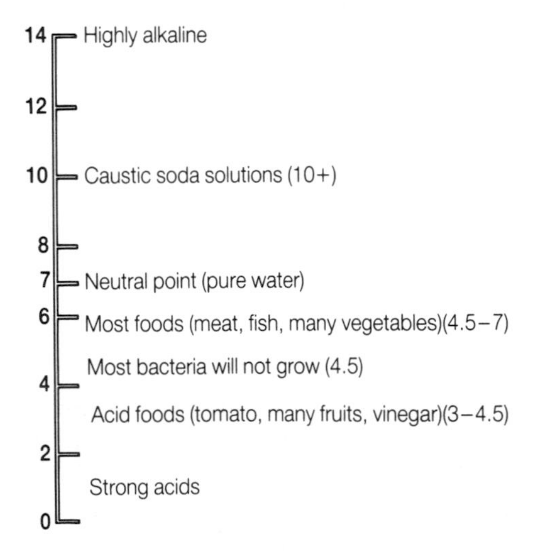

Acid/alkali pH scale

Acid foods (classification) In canning the heat treatment required depends upon several factors including the pH of the food to be processed. In general, pathogens do not grow below pH 4.5 but foods are commonly divided into several categories. These are:

- high acid (below pH 3.7, e.g. pickles and some fruits)
- acid (pH 3.7 to 4.5, e.g tomatoes and some fruits)
- medium acid (pH 4.5 to 5.0, e.g. most vegetables and mixtures such as soups)
- low acid (above pH 5.0, e.g. meats, fish, milks and some vegetables).

The more acidic the food the less heat is required during canning. *See*: Botulinum cook, Canning.

Acidophile An organism that grows best in acid conditions, where the pH is in the range 2 to 5 (-phile = lover). *See*: *Lactobacillus*.

Acidophilus milk A sour-tasting milk product in which lactose has been consumed by lactobacilli (bacteria). It is intended for people who have a lactose allergy.

Acrolein An acrid-smelling breakdown product of glycerol produced when oils and fats are heated past their smoke point. Foods cooked in such fats or oils have an unacceptable flavour.

Activity relation chart An aid in designing workplaces such as kitchens, and food production areas in which the flows of staff, equipment, raw materials and products are marked on a plan. Information from direct observation or from flow process charts is used. The plan may show up deficiencies in layout from many points of view including those of food hygiene, work study and safety. *See*: Design of food premises.

Acute

1) A disease of sudden onset, usually severe, but only lasts for a relatively short time (hours or days). An example of such a disease is food poisoning. The symptoms last only as long as the pathogens (or their products) are in the body.
2) Rapidly acting (in the case of poison for example).

See: Bait shyness.

Additives Various substances are added to foodstuffs in order to affect amongst other things colour, flavour, texture, taste and nutritional value. A large number of additives which influence food hygiene exist.

- Preservatives, such as sulphur dioxide, sodium nitrite, acetic acid and

benzoates inhibit the growth of micro-organisms and are added to a wide variety of foods.
- Antioxidants, such as ascorbic acid and tocopherols, prevent fat rancidity.
- Emulsifiers, stabilisers, colourings and others are process or aesthetic additives.

In Europe approved additives are given an 'E' number although there is some controversy concerning the safety of some of these. International agreement on additive safety is only partial and many countries have different approved lists and standards for safety testing.

Aerobe (aerobic) An organism that will grow on the surface of a solid medium exposed to air, using the free oxygen present.

Aeromonas hydrophila A Gram-negative bacterium found in both uncontaminated and contaminated fresh water and in sewage. It is able to grow at 1 °C and hence causes spoilage of food at normal refrigeration temperatures. It causes spoilage of milk and butter producing a 'fishy' flavour and it is also associated with the spoilage of eggs. An outbreak of gastroenteritis in Canada in 1986, involving raw oysters which had been stored at −72 °C for 18 months, illustrates the survival potential of this organism.

Aerosols Particles or droplets of moisture suspended in the air, invariably carrying micro-organisms. They may contaminate unprotected foods, resulting in food poisoning. Such droplets are produced by sneezing, coughing, talking, whistling etc, and particles may cause problems by dusting, sweeping floors and removing clothing. It is therefore important to protect foods from such contamination. *See*: Contamination of food, Droplet infection, Personal hygiene.

Aflatoxin An endotoxin produced by some fungi, and hence is called a mycotoxin. It was first recognised in Britain and the US during the 1960s. Several mushrooms and toadstools produce mycotoxins, as do some moulds, such as members of the genus *Aspergillus*. Some species of *Penicillium*, while growing on stored foods such as peanuts, oats, wheat, soya beans and rice produce aflatoxin. The toxin is very resistant to heat and can withstand autoclaving. If animals are fed with feed contaminated with aflatoxin they can suffer the effects of poisoning. Aflatoxins have been found in milk from cows that were fed aflatoxin-contaminated grain. There is some evidence to suggest that low levels of aflatoxin are carcinogenic to animals, especially affecting the liver.

Agar-Agar (agar) A gelling (solidifying) agent (originally used in cooking) which is extracted from a seaweed. It is now used world-wide to solidify bacteriological media, the solid 'agar gel' being used to culture bacteria in laboratories. This is often the first step in the isolation and identification of bacteria taken from, for example, a suspected contaminated food or from a food handler following an outbreak of food poisoning.

Aggressiveness A term used to describe the ability of bacteria to enter, spread and multiply inside the body of a host and so reach a susceptible tissue. It is one of the two properties (the other being toxin formation) which make some micro-organisms pathogenic. (Aggressiveness is also known as 'invasiveness'.)

Air conditioning A system of cleaning and adjusting the temperature and regulating the humidity of air used in public and commercial buildings and food stores.

Air drying

1) A method of preservation in which hot air is passed over fruits, vegetables and sometimes cereals contained in a tunnel, resulting in dehydration.
2) A method of allowing washed crockery, utensils etc. to dry naturally without the use of a cloth. This removes the possibility of recontamination of the washed items by a soiled cloth.

Albumen (egg white) In order to prevent the growth of pathogens, dehydrated albumen (a protein) is first pasteurised. Dried egg (both whole and white) has been involved in a number of outbreaks of salmonellosis world-wide. *See*: Dehydration.

Alcaligenes A group of Gram-negative bacteria which produce an alkaline medium during growth. A number of species are known to cause spoilage of foods, such as 'ropiness' in milk and a slimy growth on cottage cheese. The organisms are widely distributed in nature, being found in faeces, soil, water and dust.

Alcohol Term normally used for ethanol (ethyl alcohol), a constituent of fermented and distilled drinks. Most living organisms are not capable of tolerating high concentrations of alcohol. It has effects on the nervous system and it is also an antiseptic.

Aleukia (Alimentary toxic aleukia) A serious, often fatal disease affecting red and white blood cell functions. It is caused by a mycotoxin produced by various species of mould when millet is allowed to remain in fields over winter and harvested late. Victims suffer from severe anaemia, haemorrhaging and massive uncontrolled infections. It is not a problem if grain is harvested and stored properly. *See*: Aflatoxin.

Algae

- A large group of plants, including single-celled forms that float on water, and the seaweeds. They can cause problems by fouling drinking water and producing poisonous substances.
- They grow mainly in fresh water, but some grow in moist soil, on the bark of trees, on damp walls and floors and in sea water.
- One group found in the oceans may, particularly in the summer months, grow extensively, forming 'blooms' called red tides.They produce a strong poison which kills fish and may even kill humans. Shellfish, such as oysters or clams growing in such water concentrate the poison in their bodies which can then cause illness in people eating them.
- Annually off the north-east coast of Britain, from around May to June, shellfish become toxic by filtering the alga *Alexandrium tamarense* from the water. Such blooms are confined to relatively small areas, but the number of them reported in recent years has shown an increase in both frequency and extent. (These blooms occur more frequently in the summer months and may be one reason for the saying 'only eat oysters when there's an "r" in the month'. None of the summer months May to August contain this letter.) *See*: Blue-green algae.

Alimentary tract Alternative term for the digestive system.

Alimentary toxic aleukia (ATA) *See*: Aleukia.

Alkali A substance having a pH of over 7. Alkaline substances are rare in food, two examples being egg albumen (pH 7.5, increasing with age) and sodium bicarbonate/hydrogen carbonate (a component of baking powders). *See:* pH.

Alkali production Some bacteria (e.g. *Pseudomonas fluorescens*) spoil foods by producing alkalis. They can cause a bitter flavour in milk.

Alkaloids Organic chemical substances which will react with acids to form salts. Examples include nicotine (often used as an insecticide in horticulture), solanin (a green colour developing in the skin of potatoes stored in the light), caffeine (a stimulant occuring in coffee, tea and chocolate) and capsicin (an irritant substance found in pepper).

Allergy Consumption of certain foods by susceptible individuals may result in an allergic reaction characterised by a specific immunological response. Initial exposure produces no effect but sensitises the areas which become affected on future exposures. Many foods have been reported to produce allergic reactions of varying severity including strawberries, chocolate, milk and some food colouring materials. Allergic reactions are also possible on exposure to animals, pollen, fungal spores and many other items.

Alpha amylase An enzyme present in egg yolk and used as an index of processing for pasteurised whole egg. The enzyme is denatured by heating to 65 °C for two and a half minutes.

Alpha chloralose A rodenticide which lowers body temperature causing death by hypothermia. It is toxic to other animals and is ineffective at high ambient temperatures.

Aluminium A metal commonly used for making cooking vessels such as saucepans and also cooking foil.

Advantages

- Low cost
- Lightness
- Good thermal conductivity

Disadvantages

- Acid foods dissolve aluminium and the metal has been linked with Alzheimer's disease because of the high level of aluminium found in the brains of sufferers of this disease.
- White sauces cooked in aluminium pans may discolour due to the presence of aluminium oxide.

Aluminium sulphate is a sedimentation agent used in water treatment and was involved in an incident in Cornwall, England in 1989 causing severe and varied symptoms in a number of people who consumed accidently-contaminated water.

Alzheimer's disease Premature senility in which a region of the brain (the cerebral cortex) degenerates causing memory loss, paralysis and speech impairment. It has been linked with aluminium as a possible cause.

Ambient temperature The temperature of the surroundings, or room temperature. This may vary widely in a kitchen environment, but is often around 20 °C – well inside the danger zone for bacterial growth. Leaving foods for prolonged periods at room temperature is a common cause of bacterial food poisoning.

American cockroach (*Periplaneta americana*) A large (up to 40 mm in length) reddish-brown insect occurring mainly on ships and in port areas in Britain. The somewhat perjorative name 'Roach City' has been applied to New York due to the large numbers of these insects found there. *See*: Cockroaches.

Amino acids A group of over 20 nitrogen-containing organic substances from which proteins are made. Amino acids are involved in the Maillard reaction and contain the amino group ($-NH_2$). *See*: Browning.

Ammonia A strong smelling gas which may taint goods such as meat through accidental exposure to it. The gas is often used as a refrigerant on board ships. It is produced in over-ripe cheeses such as Brie and Camembert and it is also a sign of advanced putrefaction in meats and fish.

Amnesic shellfish poisoning (ASP) It is caused by the neurotoxic substance domoic acid produced from a marine diatom (*Nitzschia pungens*) and affecting shellfish such as mussels. Food poisoning and also short-term or permanent memory loss occurs.

Amoebic dysentery (amoebiasis) A serious illness caused by the protozoan parasite *Entamoeba histolytica*. The illness occurs following the consumption of sewage-contaminated water or food contaminated with such water. The incubation period varies from a few days to four weeks and the symptoms include diarrhoea of varying severity (usually containing blood and mucus) and death is not uncommon. The disease may be prevented by using only treated water supplies, together with effective sewage treatment. Consequently outbreaks in the UK are rare.

Ampholytic Bactericidal detergent molecules derived from amino acids and used in liquid soaps.

Amphoteric A property of surface-active agents used in cleaning chemical formulations. In alkaline conditions they are anionic (negatively charged) and possess detergent properties while in acid conditions they are cationic (positively charged) bactericides. The range of bacteria affected is restricted. Amphoteric substances may be deactivated by a number of organic substances, including food residues.

Anaerobe (anaerobic) An organism that is able to grow in the absence of oxygen. This is important with regard to food since it means that the simple exclusion of air (as in vacuum-packing) will not prevent all bacterial growth. A common food poisoning organism of this type is *Clostridium perfringens*. *See*: Canning, *Clostridium botulinum*, Sous vide.

Animals Most (if not all) wild, domestic and food animals are known intestinal and surface carriers of pathogens, parasitic worms and zoonotic diseases. For this reason live animals are not permitted in food rooms. The full extent of diseases carried by animals is very large and includes: anthrax, brucellosis, glanders, malta fever, salmonellosis, swine erysipelas, taeniasis, trichinosis, toxocariasis, tuberculosis, vaccinia and Weil's disease.

Anionic Detergent molecules having a negatively charged hydrophilic head including soaps (stearates and laurates) and soapless detergents (alkyl benzene sulphonates). Most common detergents are anionic and tend to produce large amounts of foam. *See*: Sanitiser.

Anisakis simplex A white-coloured roundworm (nematode) approx 2 cm long which is a parasite of cod, salmon and herring. Illness is caused in humans by the burrowing of the larva into the stomach and intestines.

Anthrax A disease mainly of bovine animals and sheep, also affecting humans and caused by the Gram-positive bacterium *Bacillus anthracis*, spores of which develop when the organism is exposed to oxygen. They are frequently found on pasture land. Rapid death occurs in animals with blood-stained fluid exuding from body openings and the carcass quickly decomposes. The disease is highly contagious and must be notified. Affected carcasses should not be dissected but are incinerated or buried 2 m deep in quicklime. People are infected by contact with animals, their hides or hair (weavers have contracted the disease in this way) or by eating affected meat. Human fatality is very likely if infection occurs; veterinary surgeons are in a high-risk group. *See*: Notifiable disease.

Antibiotics Chemicals (produced originally by some micro-organisms) which are able to kill or inhibit the growth of other micro-organisms. They were developed for the treatment of bacterial diseases, which is still their main use today. They have also been used in animal feeds (at low levels) to protect them from disease.

The appearance of antibiotic-resistant strains of bacteria are a serious potential health risk. Traces of antibiotics have been found in milk produced by cattle being treated for a bacterial disease. If these antibiotic-resistant organisms are also present in the milk, cheese production may be affected, due to the antibiotic suppressing the starter culture. This results in the production of second grade cheese, useful only for processing.

Antibiotics have been used as preservatives, their use in this area being strictly controlled by law.

Anticoagulants These are substances such as bromadiolone, coumarin and warfarin which stop the normal clotting of blood. They are widely used in rodenticides since their slow or chronic action does not arouse suspicion in the animal. Death results from multiple internal haemorrhages. *See*: Bait shyness.

Antimicrobial agent Any agent that is effective in killing micro-organisms or inhibiting their growth. The term is usually used to describe chemical agents, but physical agents, such as ultra violet (UV) light or very high temperatures also have antimicrobial effects. *See*: Antibiotics, Antiseptics, Disinfectant, Sanitiser.

Antimony A component of enamel used to coat metal items. Antimony is dissolved by acid foods and has been known to cause chemical food poisoning. Chipped enamelware should not be used.

Antioxidant A substance preventing oxidation reactions such as the development of rancidity in fats and oils or the browning of cut surfaces of fruits. Examples include vitamin C (ascorbic acid) and its salts, vitamin E (tocopherol), propyl gallate, BHA and BHT. *See*: Additives.

Antiseptics Chemicals that oppose infection (sepsis), usually by inhibiting the growth of micro-organisms. They may be used on the skin and other tissues, and they may be dilute disinfectants.

Antitoxin A substance produced by the body or administered into it to combat bacterial toxins such as those produced by *Clostridium botulinum* or *Staphylococcus* species.

Ants

- The garden ant (*Lasius niger*) occasionally invades food premises but it is more of a nuisance than a carrier of disease.
- Pharaoh's ant (*Monomorium pharaonis*), however, is a carrier of pathogens and nests in heated buildings such as hospitals and bakeries. It has been known to get into patients' beds.

Control measures

- All ants will eat widely, but they seem to prefer sweet or protein foods. Storage in closed containers is essential.
- Eradication in infected premises may be extremely difficult since a heavy infestation reaches all parts of the fabric of buildings and success relies on destroying the nests.

- Syrup-soaked baits may be effective as a control measure if their locations are suitable and feeding ants are periodically destroyed.
- Poison baits which are carried back to the nest or spraying with a residual insecticidal lacquer are examples of other treatments.
- Juvenile hormone has been used to control Pharaoh's ants.

Appeals It is possible for a person convicted of an offence, for example under the Food Hygiene Regulations, to appeal if they do not agree with the decision of the court. A similar procedure applies with alleged contraventions of other laws. *See*: Appendix 1: Food Law.

Appert, Nicholas *See*: Canning.

Apples Not known to cause food poisoning directly but some varieties contain sufficient malic acid to dissolve metals such as copper. Apples may be stored long-term if sound. *See*: Browning, Enzymes.

Apricot kernels Contain cyanide substances and are occasionally involved in food poisoning. Various spirits distilled from apricots have also been found to contain high levels of cyanide.

Arbroath smokies Fish, commonly whiting or haddock, preserved by hot smoking. *See*: Food preservation, Smoked food.

Argentine ant (*Iridomyrmex humilis*) A tropical species of ant found in warm buildings in temperate climates. It is an occasional invader of warm buildings including food premises. The species is aggressive towards other ants and tends to build nests in inaccessible parts of buildings making treatment difficult.

Arthropoda (Arthropods) A large group of animals including insects, ticks, lice, mites, spiders and crustacea. Some of these, such as flies and cockroaches are vectors of disease. (The name arthropod means 'jointed leg'.)

Ascarides *See*: *Ascaris lumbricoides*, Roundworms.

Ascaris lumbricoides

1) A large roundworm of pigs found in the intestines. The worm lays eggs which are excreted, then ingested by another pig and hatched in its intestines. The larvae produced penetrate the walls of the intestine and are carried in the bloodstream to the heart and lungs and then to the trachea where they are swallowed, the cycle now being complete. Livers are often badly affected with white areas called 'milk spot' and are aesthetically unfit.

2) One of the most common intestinal parasites of humans. Eggs are passed out of the body in the faeces and mature into infective larvae outside the body in soil or water. Infestation follows the ingestion of contaminated water or food, with a similar life-cycle to that in the pig. Water treatment and good personal hygiene will help prevent human infestation.

Ascorbic acid

1) Chemical name for vitamin C, a water-soluble vitamin, forming part of the intercellular cement which attaches body cells to each other.
2) The deficiency disease is scurvy which is characterised by tissue breakdown. This includes bleeding from gums and hair follicles, teeth loosening and eventually falling out, and the re-opening of old wounds and bone fractures.
3) The vitamin is destroyed by heat and by contact with oxygen and bicarbonates (such as sodium hydrogen carbonate). It is lost to a large extent in many food preservation processes such as canning and dehydration as well as during cooking.
4) Ascorbic acid is also used as an antioxidant, preventing browning in cut fruits and vegetables and as an improver in bread doughs.

Ascospore A sexual fungal spore, produced as a means of reproduction, in a sac-like structure called an ascus. There are usually two to eight ascospores in an ascus. Such spores are produced by the members of the phylum Ascomycota, members of which include the unicellular yeasts and the septate moulds, e.g. *Penicillium* and *Aspergillus*.

Aseptic The term literally means the absence of contamination by unwanted micro-organisms. In some canning and related processes (including filling cartons with milk, fruit juices etc.) the closure of the pack is performed under aseptic conditions to limit the bacterial load on the product. In practice this is difficult since it means putting a sterile product into sterile containers and closing them under sterile conditions.

Aspergillus A group of moulds (members of the fungi) which are commonly found in soil and decaying vegetation, such as in compost heaps. Many species are involved in the spoilage of foods, and some are useful in the food industry (e.g. making soy sauce from soya beans, and rice wine from the fermentation of rice). They grow well in concentrations of sugar and salt and are therefore found in many foods which have a low water content. Many members of this group produce the deadly aflatoxin.

Aspergillus flavus A mould which commonly grows on seed crops, including peanuts, oats, buckwheat, rice and soya beans. It grows best if the crop has been damaged during harvesting or if there has been insect damage in storage. The aflatoxin produced can pass up the food chain and has been found in milk (from cows and humans) and eggs.

Aspergillus niger A mould which causes spoilage of refrigerated foods, such as sausages, and is responsible for much 'mouldy' bread. It has a number of uses in the food manufacturing industry, including the manufacture of citric acid (for the soft drink industry) and it is also used to manufacture a variety of enzymes.

Aspergillus oryzae A mould used to produce sweet syrups from corn and other starches and also in Japan to produce fermentable sugars for sake (rice wine) making.

Assembly systems

1) In large scale catering operations, such as hospitals and cook-chill, a conveyor belt system of plated meal assembly is often used with a central conveyor passing between work stations where hot and cold foods are plated by operators. Temperature control using insulation or heating/chilling is required if the food is not for immediate consumption.
2) The *mise-en-place* system is widely used in which partly finished stocks are kept so that orders may be rapidly satisfied. Examples include components of sandwiches in sandwich bars, pre-prepared pizza bases and toppings, and preparation of sauces before a service period. Maintenance of temperature control is important.

See: Appendix 1: Food Law.

Asymptotic The absence of symptoms. *See*: Symptomless excreter.

ATP (adenosine triphosphate) Freshness in wet and frozen fish may be determined by the level of ATP, which is broken down, on death, into hypoxatine. The test is colorimetric. *See*: TMA.

Australian cockroach (*Periplaneta australasiae*) Similar in appearance and habits to the American cockroach but having triangular yellow markings on the front wings.

Australian spider beetle (*Ptinus tectis*) A cosmopolitan and widespread beetle found in foodstores and warehouses, which feeds on all types of food, often cereals and spices. The larvae cause damage by boring into packaging materials. The adult beetle is 5–6 mm long and covered with dense golden brown hairs. They resemble spiders because of their long legs and narrow 'waist' between the thorax and abdomen. *See*: Beetles.

Authorised officer An employee of an enforcement authority empowered to perform functions under various laws. As an example, an EHO employed by a local authority is authorised to enforce Food Hygiene Regulations and will have identification to prove this. *See*: Food Safety Act 1990.

Autoclave An instrument used in microbiology laboratories to sterilise materials such as glass, bacteriological media, water and chemicals by exposing them to moist heat under pressure. Materials are made sterile in a matter of minutes, e.g. 15 to 20 minutes at 101 to 135 kPa (15 to 20 psi). This produces a temperature inside the autoclave of 121 °C. The higher the pressure used, the higher the temperature obtained and the shorter the time required for complete sterilisation.

Autumn fly (*Musca autumnalis*) A swarming fly which enters buildings in the autumn and leaves in the spring. It lays eggs on animal dung and pupates in the soil before hatching in the summer to become a pest of horses and cattle. *See*: Flies.

Available water (a_w) A measure of the availability of water on a scale of 0.00 to 1.00 expressed as the relative humidity of the product compared with the relative humidity of the atmosphere in equilibrium. Most pathogenic bacteria grow well in a medium with an a_w of at least 0.95 although many yeasts and moulds tolerate an a_w of 0.75. Freezing at below minus 15 °C reduces the a_w of ice to around 0.85 and causes some bacteria to be killed. (*Note*: a_w is also known as Water Activity.) *See*: Growth requirements (bacterial).

Avirulent Harmless, lacking in virulence. Such organisms do not cause disease.

Avocado pear A fruit unusual in its high fat content, often over 10%. It does not ripen until picked and is very susceptible to chilling injury during storage and enzymic browning when cut.

a_w *See*: Available water.

Bach process A heat-sealing process in which thermoformed plastic lids are applied to containers of food such as yoghurt.

Bacillary (bacterial) dysentery A diarrhoeal disease more common in tropical countries than in Britain, particularly affecting infants and young children.

- Incubation period is from two to four days.
- Symptoms include diarrhoea, abdominal cramps, pain, fever, vomiting and, if severe, blood and mucus or pus in the faeces.
- Caused by members of a group of Gram-negative bacteria called the *Shigella*, and is also known as shigellosis.
- Commonly spread by poor personal hygiene (i.e. faeces-contaminated hands), inanimate objects (such as toilet seats), food, water and flies.
- Treatment requires antibiotics and the essential replacement of body fluids lost as a result of diarrhoea and vomiting.

Sanitary procedures, including effective sewage treatment, water purification, control of flies, together with sanitary preparation and serving of food all reduce transmission of the disease. The *Shigella* are the second most frequent cause of 'traveller's diarrhoea'.

Bacillus

1) A term used, in the lower case ('bacillus'), to describe the shape of certain bacteria, which are rod-shaped. *See*: Bacterial classification.
2) When written as '*Bacillus*', it is the name of a specific group (genus) of bacteria. *See*: *Bacillus cereus*, *Bacillus subtilis*.

Bacillus cereus This organism is widely distributed in nature, where it is found in the soil, cereals, vegetables, dairy products and many more places.

- The onset period for the illness is short, from 1 to 16 hours and lasts for 12 to 48 hours.
- Symptoms are similar to those for *Staphylococcus aureus* food poisoning and appear suddenly, with acute vomiting and some diarrhoea.
- Foods affected are notably cereals, such as cooked rice (particularly when boiled but eaten cold in salads); fried rice (especially re-heated); and cornflour.
- Cooked rice must be refrigerated until used to prevent germination of spores.

It is a Gram-positive, aerobic, spore-forming, bacterium which causes food poisoning due to the release of an exotoxin. *See*: Endospore, Enterotoxin.

Bacillus stearothermophilis A heat-resistant organism (thermophile) that causes the development of acids in canned foods resulting in a spoilage condition known as 'flat-sour'. It is an anaerobic spore-former found in low-acid foods such as vegetables and milk. An Fo value in milk for this bacterium would be in the order of 5–6.

Bacillus subtilis This is a Gram-positive aerobic to facultative spore-forming bacterium. It can cause food poisoning, with an onset period from 2 to 18 hours. Symptoms include vomiting, diarrhoea and abdominal pain. It usually results from a chance contamination, and is not associated with any particular food. It is also responsible for spoilage of foods, such as canned cured meats and causes 'ropiness' in bread. It is also used in some biotechnology processes, to produce enzymes in 'biological' washing powders, as well as in the production of antibiotics.

Bacon A pork product produced by curing which kills most bacteria and parasites such as tapeworms and roundworms. Certain Staphylococcus species survive the high salt concentration (they are called halophilic) and may cause food poisoning if temperature control is not adequate. Bacon should be cooked thoroughly before consumption. *See*: Pork.

Bacon beetle (*Dermestes lardarius*) A pest of skin, hide, biscuits and protein foods. Remains of dermestid beetles have been found in mummified bodies and the beetle is known to feed on dead rodents and bonemeal.

Bacteria (singular Bacterium),

- There are over 1 700 known species of bacteria, only some of which are pathogenic.
- The vast majority are free-living and play useful roles in nature, including:
  - a) the recycling of materials via the processes of the decomposition of organic matter, as in soil;
  - b) the treatment of sewage;
  - c) soil improvement by the production of nitrites and nitrates.

People have also been using bacteria for hundreds of years in the production of food, such as cheese and yoghurt. A few forms are photosynthetic. In recent years bacteria have been increasingly used in industrial processes (biotechnology), for example in the production of drugs. However, some species are pathogenic (i.e. cause illness) while others cause spoilage and decomposition of food.

Bacterial cell (structure) Not all bacteria possess a capsule (also known as a slime layer). Similarly only some forms possess flagella, which are used for locomotion. All forms possess fimbriae. The bacterial cell has a much simpler structure when compared to the typical animal or plant cell; the bacterial cell is described as prokaryotic, and the typical animal or plant cell is described as eukaryotic.

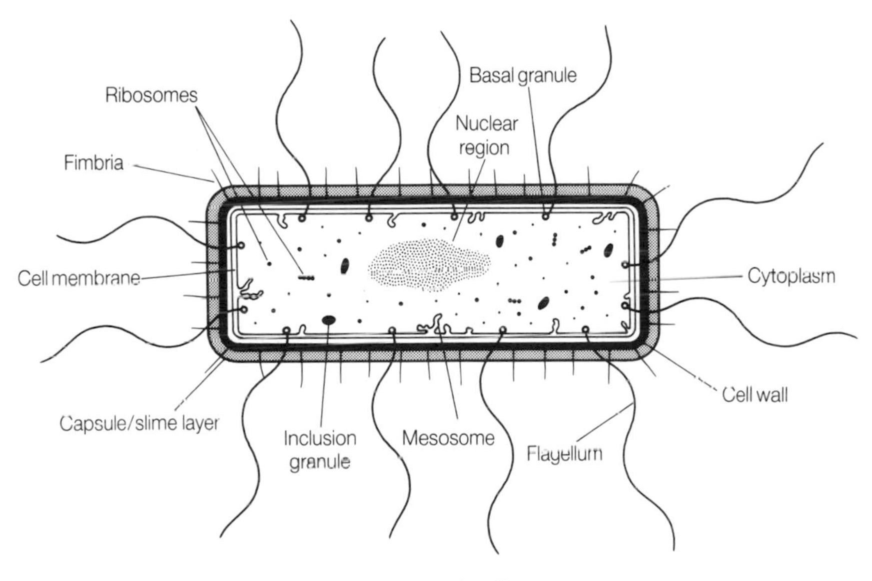

Bacterial cell

Bacterial classification

- Bacteria are generally unicellular (single-celled), often existing in groups or aggregates of cells.
- Individual cells are extremely small, and most are no larger than 1 to 2 micrometres in diameter, and 1 to 10 micrometres in length. (1 micrometre = 1 millionth of a metre.) One simple means of classifying bacteria is based on the shape of the cell. The main groups are:
  1) the coccus group (pl. cocci), which are spherical
  2) the bacillus group (pl. bacilli), which are rod-shaped
  3) the vibrio, which are comma-shaped
  4) the spirillum group , which are spiral or corkscrew-shaped.

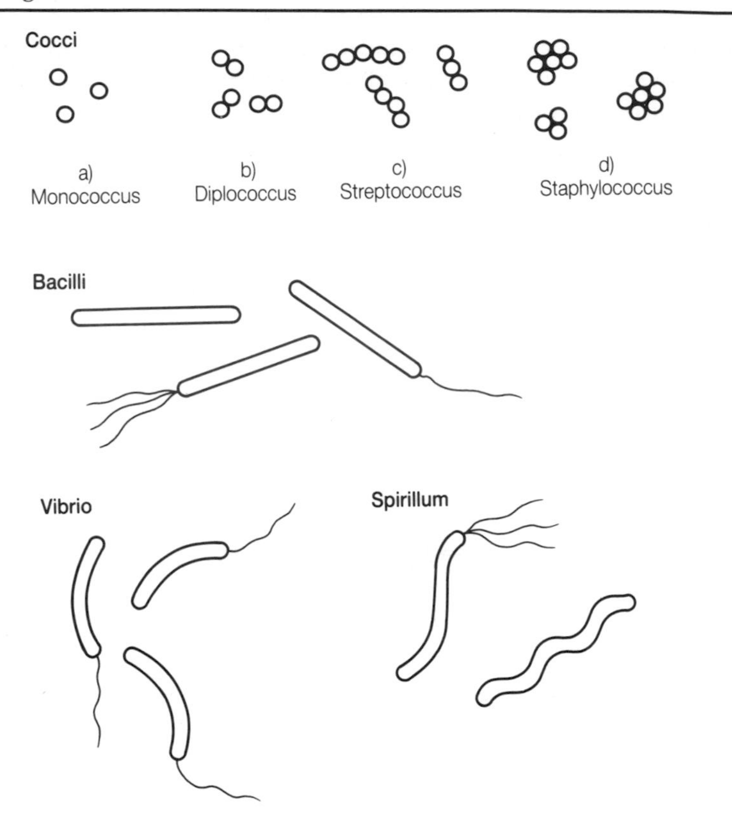

Bacterial shapes

Of the four groups, the coccus group may be further sub-divided into colonial forms, the cells existing in groups or clusters. The shape of the cell is used to help in identifying bacteria. Another aid to identification is a special staining procedure, called the Gram stain. *See*: Diplococcus, Staphylococcus, Streptococcus.

Bacterial growth Bacteria are living organisms and so have the usual requirements for growth.

- Food (nutrients)
- Warmth (specific temperature requirements)
- Moisture (water activity)
- A certain gaseous environment such as oxygen
- A certain pH
- Time

During periods of growth and reproduction, usually by binary fission, the cell is described as vegetative. A doubling of numbers of bacteria may occur in ten to twenty minutes under ideal conditions.

- When conditions become unfavourable for growth, some forms of bacteria are able to enter a resistant phase known as an endospore or spore.
- This enables them to survive under adverse conditions, such as the very high temperatures obtained during cooking.
- If and when conditions again become favourable, they return to the vegetative phase and continue to grow and reproduce as before.
- Bacteria produce and release chemical poisons called toxins and it this these that are responsible for the symptoms of many diseases, including food poisoning.

Note: Bacterial endospores are *not* produced as a result of reproduction, which is the case with fungi. *See*: Binary fission, Endotoxins, Enterotoxins, Exotoxins, Growth requirements (bacterial), Spores (fungal).

Bacterial growth curve When bacteria are grown in a closed system (one which contains all the necessary growth requirements; nothing being added or removed) in liquid growth medium, and the population is counted at intervals, it is possible to plot a growth curve. Such a curve shows four typical phases of growth over a period of time.

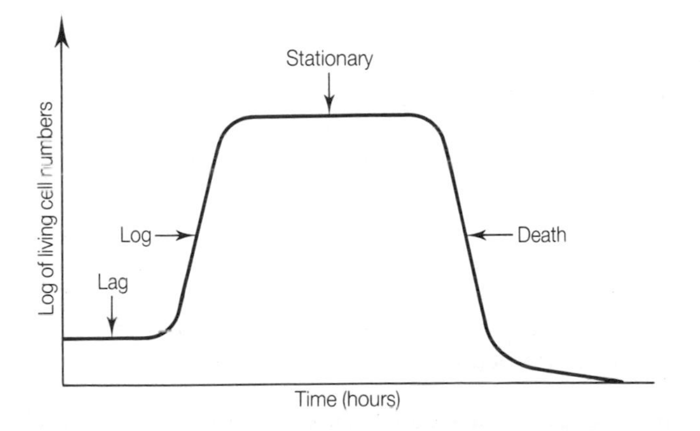

Bacterial growth curve

1) **The lag phase.** During this phase the bacteria are adapting to the new environment. It can last from about one hour to several days.

There is some reproduction, but the number does not increase since the reproduction rate is equal to the death rate. The end of the lag phase and the beginning of the next phase occurs when the bacteria have made the necessary adaptations.

2) **The log phase.** Reproduction is occurring at the maximum rate. The bacterial cells are reproducing exponentially and there is minimum cell death. As nutrients become scarce and the wastes accumulate, there is a decrease in the growth rate and the curve turns into the next phase.
3) **The stationary phase.** Growth and reproduction slow considerably to a point where the reproduction rate equals the death rate. This is often due to the loss of a growth factor, such as a vitamin or amino acid, or a sudden change in pH. This eventually leads to the final phase.
4) **The death or decline phase.** This is the reverse of the log phase, where death of cells occurs at an exponential rate. Death rate greatly exceeds reproduction rate. Accumulation of wastes is a common cause. Eventually the whole population of cells will die out.

Bacterial growth curve (Implications) Knowing how a population of cells increases through time has practical implications in the food industry.

- An awareness of the bacterial growth curve may be used to explain the spoilage of certain foods, such as milk. Pasteurised milk may be kept in a refrigerator for a few days only before it spoils, even if unopened. The spoilage agents are those organisms in the milk which are unaffected by the pasteurisation process. During unopened storage or after opening the spoilage organisms present enter the lag phase. Each time the milk is taken from the refrigerator the temperature increases sufficiently to allow the micro-organisms to adapt to their environment. The longer it is left outside the refrigerator, the quicker the lag phase will proceed. Once the lag phase is completed the milk may spoil in three to four hours. Continued refrigeration will extend the lag phase, but it will finish eventually and the milk will be spoiled.

- If bacteria are given ideal conditions, such as in a warm kitchen or food production environment, there may be a rapid increase in bacterial cell numbers, as the bacteria enter the log phase, and this may result in an outbreak of food poisoning. The easiest way to stop this is to use proper temperature storage of foods. *See*: Temperature control of foods.

Bactericidal soaps (and lotions) Preparations containing bactericidal substances, designed to reduce the numbers of bacteria (particularly *Staphylococci*) on the hands. In order to be effective they require persistent and consistent use.

Bactericide An agent designed to kill bacteria, such as a disinfectant. ('-cide' means 'killer' or 'killing'.)

Bacteriological standard The quality, acceptability or potential safe shelf-life of foods may be measured in part by the numbers of micro-organisms present. There are several available laboratory methods including:

- **Total viable count (all living bacteria)** Numbers greater than 10^4 organisms/gram indicate an unacceptable level of contamination.
- **Faecal organisms** An indication that contamination from intestines of humans or animals is present.
- **Specific pathogens** *Staphylococci*, *B. cereus*, *C. perfringens* and *Salmonellae* among others may be tested for specifically to highlight any problems such as poor personal hygiene or cross-contamination.

As an indication of levels, *Salmonella* species should be absent, *C. perfringens* is relatively safe at 10^3/gram, *S. aureus* at 10^2/gram and *E. coli* at 100/gram. These figures are intended as a guide only.

Foods may be tested as part of a quality control system such as HACCP or after samples are taken by an EHO. In Britain there are no legal standards at present except for milk, although there is power in the Food Safety Act 1990 to make regulations concerning the levels of bacterial contamination in foods. *See*: Milk.

Bacteriophage (phage) A type of virus that infects bacterial cells.

Bacteriophage (phage) typing A method of identifying bacteria using specific strains of bacteriophage. An example is *Salmonella enteritidis phage type 4*, the organism associated with causing food poisoning following consumption of raw or lightly cooked eggs. They are highly specialised in that they attack only members of a particular species or even strains within a species.

Bacteriostatic Halting the continued growth of bacteria but not necessarily reducing their numbers. Substances such as antiseptics used on the skin possess this property.

Bain Marie

1) Equipment for keeping food hot, consisting of metal containers, with or without lids, suspended over hot water in a cabinet. There is also a dry, mobile version heated by gas or electricity. Construction is usually of stainless steel. Food should be capable of being held above 63 °C, out of the danger zone.
2) A heated water bath used to keep containers of sauces, for example, hot but not boiling, or to cook these gently.

Baits Substances placed in traps to attract food pests. They may be poisonous to kill pests or non-poisonous to estimate the scale of an infestation. A rodenticide bait may contain a mixture of poison and cereals, meat, fish or other food. Baits are not effective if there are other food supplies available. Bait boxes must be numbered, marked as being poisonous and placed at known locations marked on a plan of the premises.

Bait shyness Refusal of rodents to eat further bait if earlier consumption of the bait has caused illness but not death. Another bait may be substituted.

Bakeries Produce a wide range of foods such as breads, cakes and pies some of which are high-risk foods containing meat, eggs and cream. Warm conditions around ovens have provided ideal living conditions for cockroaches and stored cereals are sometimes infested by various beetles.

Bamboo shoots Natural, non-cultivated varieties of bamboo may contain cyanogens which release hydrogen cyanide by enzyme action when the plant cells are damaged. *See*: Cyanide.

Bandages Required by law in first aid kits for use in food rooms. The numbers and types of bandages depends on the number of people employed and the risks of the work. Under Food Hygiene Regulations a bandage is not a 'suitable waterproof dressing' and should not be worn where it could come into contact with food. A finger bandage could be satisfactory if protected by a finger stall which would prevent any bacteria from an injury from getting onto food.

Barbecues The main risk is that foods may be cooked on the outside but fail to reach an adequate internal or centre temperature. This is a particular problem with 'made up' foods such as beefburgers since any bacteria present will be spread throughout the product. Undercooked chicken may be a hazard owing to the probable survival of *Salmonella* organisms.

Barrier cream Protects the skin of the hands from drying and cracking after repeated hand-washing. The cream may also be bactericidal. Food grade barrier cream should not cause smell, taste or other taint in food.

Basal body *See*: Flagella.

Basidiospore A sexual fungal spore, produced as a means of reproduction, formed externally on a club-shaped pedestal structure called a basidium. Such spores are produced by mushrooms and puffballs.

Basins *See*: Handwashing.

Beans Many beans (such as red kidney beans) and pulses are known to contain toxic chemicals and must be thoroughly cooked before consumption. *See*: Red kidney beans.

Beef

- Meats may be contaminated with bacteria by cross-contamination at the time of slaughter or during food delivery, storage and preparation.
- The spores of bacteria such as *Clostridium perfringens* may be present on the surface of beef having come from the soil via the animals' feet and hide.
- The insides of joints or pieces of meat normally contain few bacteria and rare beef, unless minced, poses only a slight health risk with tapeworm.
- Cooked beef is a high-risk food and must be refrigerated to below 8 °C or kept hot above 63 °C. *See*: Appendix 1: Food Law.

Beefburgers Since any bacteria on the surface of beef is mixed throughout a beefburger, they must be thoroughly cooked so that any organisms in the centre are destroyed. An outbreak of food poisoning involving *E. coli* occured in Britain in 1991 due to the undesirable practice of serving rare beefburgers.

Beef tapeworm (*Taenia saginata*)

- The cysts (larvae) of this inhabit beef muscle, particularly those with a high blood supply, such as the heart, diaphragm and cheek.
- These are difficult to detect since they are small (pea sized at most) pearly-white objects embedded in muscle.
- Humans are infected by eating undercooked meat.
- The tapeworm, which can be up to 15 metres long, develops from the cysts in human intestines and attaches itself to the walls of the intestine with hooks/suckers which may tear the intestine, causing it to bleed.
- Segments containing eggs are periodically released and pass out with the

faeces. They may contaminate grazing land, starting the cycle once more.
- Cysts are destroyed by thorough cooking or freezing at −10 °C for at least 14 days.
- The cysts are called cysticercus bovis, cattle being the intermediate host.
- Around 60 million cases per year world-wide are estimated. Treatment for humans is now safe and effective.

Beetles (Coleoptera) Many varieties infest food, causing spoilage but not food poisoning. Affected food is unfit for human consumption. Infestations are often difficult to deal with and fumigation with methyl bromide or other substances may be the only solution if an entire building has become infested. There are very many beetles, almost one for every different food, including Australian spider, Bacon, Biscuit, Broad-horned flour, Cadelle, Cigarette, Coffee bean, Confused flour, Copra, Dried fruit, Flour, Golden spider, Khapra, Long-headed flour, Maise, Mealworm, Merchant, Pulse, Rice, Rust-red flour, Rust-red grain, Saw-toothed grain, Small-eyed flour and Spider beetles.

Benzene hexachloride (BHC) A persistent insecticide also known as HCH (hexachlor cyclohexane). The gamma isomer is highly insecticidal and has been used against a wide range of insect pests. It is formulated in sprays, dusts and powders, ointments, smoke generators and as brickettes. It acts against the central nervous system of insects and also affects mammals.

Benzoic acid A preservative used to prevent mould and yeast growth in fruit juices, pickles, soft drinks, coffee essence, salad cream, beers and other foods. Sodium benzoate, calcium benzoate, potassium benzoate and ethyl 4-hydroxybenzoate have similar uses. These substances may cause adverse reactions in asthma sufferers and persons having skin allergies.

Besatz Impurities such as weed seeds, fungi, rodent hairs, dust, stones and other physical contaminants in wheat grain received at the mill. This European term is called 'dockage' in the USA.

'Best-before' date A coding system used by food manufacturers to indicate the time during which a product is at its optimum quality. Typical foods have long shelf-lives of months or years and are low-risk foods.

- Includes bakery goods such as cakes and biscuits, canned foods and drinks, dried foods and crisps.
- There is little risk of food poisoning from them.
- It is a quality statement; the eating quality of the food deteriorates slowly after the date.

High-risk foods should have a 'use-by' date since they are perishable and may cause food poisoning.

BHA and BHT Synthetic antioxidants, butylated hydroxy anisole and toluene, used to prevent oxidative rancidity in foodstuffs. The safety of both substances has been in question and there are restrictions on their use in infant foods although BHT has been sold as an antiviral agent and BHA is reputed to protect against some carcinogens.

Biguanide A detergent/bactericide (sanitiser) used for glass washing.

Binary fission

- A process of cell division occuring in a number of single-celled organisms, including the bacteria, in which one cell divides into two daughter cells.
- Each daughter cell produced is similarly able to divide, resulting in 4 cells, then 8, 16, 32, 64, 128 and so on.
- Bacteria are able to reproduce in this way in as little as ten to twenty minutes, and will continue doing so as long as their surroundings remain favourable.

Binary fission results in large numbers of cells being produced in a fairly short period of time. *See*: Bacterial growth curve.

Biodegradability The ability of a substance to be broken down into non-toxic or non-polluting end-products after its use. Many plastics are non-biodegradable as is the chemical disinfectant hypochlorite. Hydrogen peroxide, a bactericidal chemical, breaks down forming oxygen and water and is biodegradable.

Biotechnology The application of micro-organisms, their processes or products in industry. Historically, micro-organisms such as yeasts and bacteria have been used in the production of alcoholic beverages, cheesemaking and baking. More modern uses include the treatment of sewage and the production of drugs.

Biphenyl An antifungal preservative used on the outsides of fresh fruit, especially citrus fruits. Washing with detergent helps remove it.

Birds Pests of food rooms, particularly sparrows and pigeons.

Problems

- They carry food poisoning bacteria such as *Salmonella* and also carry insect pests.
- Droppings and feathers may cause physical contamination of food.

- Birds deface buildings and nests may create a fire risk.
- The diseases psittacosis and ornithosis, caused by unusual organisms called the *Bedsonia*, are carried by many species including parrots and budgerigars, causing fevers and pneumonia in humans.

Control
- Buildings may be proofed against entry by fixing screens or netting over doors and windows.
- External window ledges and similar sites may be sloped or treated with soft materials to discourage perching and nesting.
- Narcotising with alpha chloralose is a possible control measure, as are shooting, trapping and using scaring devices.

See: Zoonoses.

Birdseye, Clarence An American pioneer of freezing who developed the commercial operation of this preservation method during the 1920s. *See*: Freezing.

Biscuit beetle (*Stegobium paniceum*) A pest of stored cereal products. It is small, reddish-brown in colour and covered with yellow hairs. It is the most common insect pest of domestic food stores. In the USA it is known as the Drug store beetle.

Biscuits Many are both low in water and high in sugar and are thus not capable of supporting the growth of bacteria. They are however attractive foods for rodents and insect pests and should be stored in containers with tightly-fitting lids at low humidity (RH 60%) at between 10–15 °C. *See*: Dry goods store.

Black ant (*Lasius niger*) It is also called the Garden ant and sometimes invades food rooms. Unlike Pharaoh's ant it is not known to carry disease and is more of a nuisance although infested food should be regarded as unfit. Infestations may be persistent and it is important to practise good housekeeping so that food is not available. Destruction of ants is by insecticidal dusts and sprays which must be used carefully in food areas.

Black bat, Black beetle, Black clock Alternative common names for the Oriental cockroach, *Blatta orientalis*.

Black Death Another name for the Great Plague (bubonic plague) that swept across Europe in the fourteenth century. It spread from rat to human by the rat flea (*Xenopsylla cheopis*) resulting in over 25 million deaths (about one third of the population). The causal organism is a bacterium, *Yersinia pestis*. Epidemics are rare but still occur today. The reservoir of infection is wild rodents that are resistant to the disease.

Black fungus beetle (*Alphitobius laevigatus*) A black oval beetle about 5 mm long breeding in damp stored cereal grain.

Black pudding A cooked meat product made from blood, oats, cubes of fat and seasonings encased in a skin of animal intestine. It should be refrigerated during storage as it is a high-risk food.

Black rat (*Rattus rattus*) The least common of the two main species found in the UK. It is also called the Ship's rat and is rarely found away from sea ports. It is smaller and more slender than the brown rat and climbs easily throughout buildings. Distinguishing features include a dark brown/black body, large ears, pointed snout and a tail longer than the body. *See*: Rats.

Black spot Surface spoilage of meats caused by moulds, usually (but not exclusively) by *Cladosporium herbarum*. The spots are usually localised.

Bladderworm The larva (also known as a cyst or cysticercus) of a tapeworm, which is located in the skeletal muscles of e.g. a cow or pig. If the meat of an infected animal is insufficiently cooked and eaten, the surviving larvae will attach to the lining of the intestine and develop into mature adult tapeworms. *See*: Tapeworm.

Blanching Treatment of vegetables by placing in boiling water or steam for less than a minute prior to freezing. The heat of the process destroys spoilage enzymes which would otherwise cause browning, loss of flavour and development of off-odours during frozen storage. Enzymes which destroy vitamins A and C may also be destroyed. Other effects of blanching include driving out some trapped air and destruction of surface bacteria. Over-blanching may result in the cooked product losing firmness of texture. *See*: Browning.

Blast chiller

- A fan-assisted refrigeration cabinet designed to rapidly lower the temperature of hot food to below 5 °C prior to refrigerated storage.
- The maximum time for this should not be more than 90 minutes to minimise the amount of time cooked food spends in the danger zone.
- The size and thickness of food has a great effect on chilling time. Joints of meat that are to be chilled should be a regular shape and no heavier than 2.5 kg. *See*: Cook-chill.

Blast freezer Similar to the blast chiller but hot cooked food is reduced to minus 18–22 °C prior to freezer storage. Quick freezing is essential to preserve food quality. *See*: Cook-freeze, Freezing.

Bloaters Whole ungutted herring preserved by dry salting and cold smoking. *See*: Curing.

Bloom The desirable surface appearance in fresh meat due to the reaction of the protein myoglobin with oxygen. Further oxidation produces metmyoglobin, a brownish grey substance indicative of increased age of meat.

Blowflies (*Calliphora, Lucillia, Phormia* and *Sarcophaga*)

- Large flies about 10 mm long breeding in raw and decaying meat and foods of animal origin.
- Some breed on live animals, faeces and decomposing plants. They are most active in warm, sunny weather.
- Eggs are laid, hatching within 24 hours into larvae or maggots. These burrow into the affected food while feeding on it.
- After some days the maggots migrate, often at night, into the soil and within a week of hatching form the pupae. The emerging flies are able to lay eggs.

Blowflies are known carriers of *Salmonella* and *Clostridia*. Affected food is termed 'fly-blown'. Alternative names are bluebottle, greenbottle and grey flesh fly. *See*: Flies.

Blown cans Cans in which gas has been produced during storage develop protruding ends. These are termed 'blown', and the contents are unfit to eat. Some bacteria of the *Clostridium* and *Bacillus* groups are heat resistant (thermoduric) and will survive if the can is underprocessed, later producing gas during storage.

Hydrogen and carbon dioxide gases cause various stages:

- 'flipper' – the can end may be forced back with light thumb pressure but flips out if the can is struck
- 'springer' – pressure on a protruding end causes the other end to spring out
- 'hard swell' or 'blown' – both ends permanently protrude.

Blown cans are most common if cans are stored at temperatures above 15 °C, or if underprocessing occurs.

Bluebottle Alternative name for blowfly. *See*: Flies.

Blue-green algae (*Cyanobacteria*) A group of photosynthetic micro-organisms which grow as a slimy or jelly-like growth on soil or on the surface of water. Under certain conditions they may grow extensively in water, forming scums or blooms. This usually occurs in late summer or early autumn. The algae will ultimately die and will be decayed by other aquatic micro-organisms. They may release toxins, which poison fish and other animals; sheep and dogs have died in Britain through drinking such contaminated reservoir water. The toxin most often implicated in animal deaths is known as microcystin, which is produced by the alga *Microcystis*. The toxin causes liver damage and people drinking such contaminated water, even accidently during swimming are also at risk. Direct contact with the alga may cause eye and skin irritations.

Body blank A piece of tin-plated steel from which the body of a can is formed.

Body fluids All body secretions and excretions are likely to contain bacteria, including many types which are pathogenic. These fluids, which include saliva, blood, pus, urine and faeces, may, if allowed to contaminate foods, give rise to an outbreak of bacterial food poisoning. The potential dangers associated with this from human sources have formed the basis of the need for strict personal hygiene. (*Note*: Body fluids from any animal present similar risks.)

Boils Infections of hair follicles by bacteria such as *Staphlyococci* resulting in a painful red swelling with a yellowish top or head.

Problems
- The head contains pus, a mixture of bacteria, toxins and white blood cells, which is highly infective. Pus must not be allowed to come into contact with food.
- In severe cases, food handlers are excluded from work until the condition has cleared. *See*: Septic lesions.

Bone taint Off-odours caused by putrefying bacteria in deep-seated parts of muscle near to the bone . The bacteria gain access from lymph nodes in the surrounding tissue and multiply rapidly if pH and temperature are high.

Booklice (*Psocids*) Small or minute wingless insects varying in colour from grey to brown, often yellowish, and infesting a wide range of stored products, mainly packaged cereals. Other foods and packaging materials may also be affected. They thrive in warm and damp conditions feeding on mildew and mould and are not known to carry pathogens. Dry conditions are the best defence against booklice. Affected food is unfit.

Boric acid A long-lasting, slow-acting insecticide used to kill ants and cockroaches. It is often applied at high concentrations and may cause damage to humans and animals if inhaled or ingested.

Botrytis A mould affecting stored food products. *Botrytis cinerea* is also called 'noble rot' and is a desirable condition affecting grapes used to produce various sweet wines in Sauternes and Rheingau. *See*: Moulds, Mycotoxins.

Bottling A preservation process similar to canning in which food is heated in sealed containers to destroy pathogens. Home bottling of vegetables has sometimes resulted in outbreaks of botulism particularly in the USA. The temperature reached by domestic equipment is not sufficient to destroy spores of the bacterium *Clostridium botulinum* which will later germinate and produce toxin. *See*: Botulism, Botulinum cook, Canning.

Botulinum cook In the canning industry, the bacterium *Clostridium botulinum*, being very heat resistant, is used as a process indicator for foods such as meats and vegetables which have a pH above 4.5. If it is destroyed then it is assumed that other pathogens are also killed and that commercial sterility has been achieved. *See*: Canning, Fo value.

Botulism A disease caused by the toxin of the Gram-positive bacterium *Clostridium botulinum*, several types of which affect the central nervous system.

- First symptoms normally occur within 18–36 hours of eating the affected food and include tiredness, headache and diarrhoea.
- Difficulty in speech, loss of balance, constipation and double vision follow within a few days.
- The disease is often fatal since the brain centre controlling breathing is affected.
- Diagnosis may not occur since the disease is uncommon and symptoms are sometimes unclear.
- Recovery of survivors is slow and may be incomplete.

The disease has been caused by home bottled vegetables (asparagus, green beans and others), canned fish (tuna and salmon), duck pate and hazlenut yoghurt but is relatively rare. Because of the severity of symptoms, cases often attract great publicity. Less than 100 cases per year are reported annually in the USA and far less elsewhere.

Bowel The gut or small and large intestines.

Brachonids Small or minute black insects (wasps) which breed inside other insect pests, their presence indicating a large population of these. *See*: *Ichneumon* flies.

Brawn A meat product made from pig's head meat, ox cheeks and beef and pork trimmings. Diced meat is cooked and moulded with gelatine or gelatinous stock. It is normally served cold and since it is a high-risk food, it should be refrigerated during storage. *See*: Meat products.

Bread Not a well-known cause of food poisoning, its water content being too low for bacterial growth. Oven temperatures will kill any bacteria present in flour although survival of spores of *Bacillus subtilis* and others may result in the condition called 'rope', causing an unpleasant odour and sticky yellowish-brown spots in affected bread. Sandwiches and similar bread products which contain high-risk foods must be refrigerated if kept for more than four hours. *See*: Appendix 1: Food Law, Moulds, Mycotoxins.

Brine A solution of salt (25%), potassium or sodium nitrate (3%) and sometimes sugar (up to 1%) in water used in curing. Also called 'pickle'.

Brining The immersing of prepared sides, pieces or slices of pork in brine during the production of bacon by curing. Brining of vegetables prior to pickling to reduce their water content which would otherwise dilute the vinegar is also practised. *See*: Sauerkraut.

British Food Manufacturing Industries Research Association (BFMIRA) A research laboratory providing various services such as food analysis to the food industry.

British Pest Control Association (BPCA) The professional body of pest control operators which lays down standards for this industry. Members must conform to the Association's Codes of Practice.

Broad-horned flour beetle (*Gnathocerus cornutus*) A brown/black beetle about 4 mm long having very large lower jaws (mandibles) and infecting food stores, particularly if food has been spilled.

Broad Street pump In 1854 the public water pump in Broad Street (now Broadwick Street) in the Soho district of London, England, was the centre of a major outbreak of cholera. The epidemiologist John Snow (later knighted) linked the sewage-contaminated water supply with the outbreak.

Brodifacoum An anticoagulant rodenticide causing death by prevention of blood clotting. Although classed as a chronic poison it may be effective after a single feed. *See*: Warfarin.

Bromadiolone An anticoagulant chronic rodenticide often formulated as a paste and extruded from a pressure gun. *See*: Warfarin.

Brown FK 154 A mixture of azo dyes used to colour kippers. Researchers have noted that this dye causes genetic mutation in bacteria.

Brown house moth (*Hoffmannophila pseudospretella*) Occasionally a pest of stored products causing spoilage of food but not carrying disease. Cereal grains especially if damaged are prone to attack. Affected food is unfit. *See*: Moths.

Browning Food spoilage may occur by browning reactions in the cut or damaged surfaces of fruits and vegetables. The reaction is promoted by oxidase enzymes released when cells are damaged and the resultant brown colour is caused by oxidation of phenols in the cell. Several methods including blanching, immersion in salted water, treating surfaces with lemon juice or other weak acids and the use of preservatives such as sodium metabisulphite are used to prevent browning. The enzymes are inhibited by heat, low pH, ascorbic acid (vitamin C), sulphite and chloride ions while immersion in water limits oxygen availability. *See*: Blanching, Bruising.

Brown rat (*Rattus norvegicus*) The commoner of the two main species, it is found throughout the UK. It is larger than the black rat, having a blunt snout, small ears and a tail shorter than its body. Body colour is brown/grey with a pale belly. It is capable of climbing but is not as agile as the black rat.

Brucella abortus A Gram-negative bacterium causing brucellosis or contagious abortion in cattle and undulant fever in humans.

- It causes a severe and sometimes fatal fever which may recur periodically for years in humans after infection from infected milk or milk products or contact with affected animals.
- It occurs mainly in dairy farmers, the veterinary profession, and slaughterhouse workers, with the first symptoms not appearing for several months after infection.
- The disease is difficult to cure with sufferers having sudden pains, bronchitis, high temperature and sweats.

The word ‘undulant’ means ‘wavelike’ and illustrates the irregular and periodic nature of the disease. It is also called Malta fever, Cyprus fever and Gibraltar fever due to its long history in Mediterranean regions where it was investigated by David Bruce, an army surgeon in the early 1900s. Pasteurisation kills the organism and infected animals are slaughtered.

Brucella melitensis A Gram-negative bacterium which causes a condition in goats similar to that caused by *Brucella abortus* in cattle.

Brucellosis *See*: *Brucella abortus*.

Bruchus beetles Pests of dried pulses, peas and lentils which lay eggs on developing pods. Larvae hatch out and feed on dried stored beans, etc. Affected food is unfit. *Bruchus ervi* affects lentils and *Bruchus pisorum* affects dried peas.

Bruising A cause of enzymic browning in fruits and vegetables when cell enzymes are released. Severe bruising causes decay to speed up and food to become rapidly unfit to eat. *See*: Browning, Enzymes.

Brush strips Fixed to the bottom edges of doors in proofing of buildings against entry of pests.

BSE Bovine spongiform encephalopathy or 'Mad Cow Disease' is caused by a virus and results in a brain disorder in cattle.

- Loss of balance and difficulty in muscle coordination are seen as symptoms.
- Originally a disease of sheep, the condition has reputedly spread to cattle via animal feed containing infected tissues of other animals.
- Animal feed may contain animal tissue as a protein supplement.
- Concern has been expressed that the condition may spread to humans; it has also been noted in pigs and other animals.
- Movement and export of beef have sometimes been prevented as a control measure.
- Improved slaughterhouse hygiene in which care is taken not to allow particles of the brain or spinal cord of affected animals to cross-contaminate other carcases has also been advised.

Buckie (*Buccinium undatum*) A mollusc about 10 cm in diameter having a pale coloured thick spiral shell containing dark coloured flesh. Buckie is an alternative name for whelk. *See*: Shellfish.

Buckling Salted hot smoked herring fillets. *See*: Curing, Smoking.

Buffers Chemical additives which resist changes in pH of foodstuffs. Many salts of citric, tartaric, malic and phosphoric acids have wide food industry uses as buffers.

Buffets Food presented on open display, often self-service.

- Cases of food poisoning have resulted from food being left for too long.
- Contamination during service is very likely.
- High-risk foods may be left at room (ambient) temperature for buffet service for up to four hours, but must be of good quality beforehand.
- Refrigerated displays for cold buffets are preferred.
- Topping up with fresh food is undesirable since the previous food may be contaminated. *See*: Appendix 1: Food Law.

Building regulations Comprehensive regulations aimed at making premises safe for use and covering every aspect of the construction of all types of building including food premises. Inspections are made by the local authority at several stages during building construction to ensure that the legal standards are being met.

Buildings (proofing of) In order to prevent rodents, birds and insects gaining access to buildings many proofing measures are possible. These include:

- fitting wire cages to the tops of ventilation pipes
- ensuring that doors, etc. fit well
- placing fine mesh screens over opening windows.

Gaps around pipes passing through walls, broken drains, defective water seals in gullies, sinks and WCs and air bricks will also allow entrance to pests as will more obvious holes in walls and ceilings.

Bulgarian bacillus Original name given by Metchnikoff in 1910 for the Gram-positive bacterium *Lactobacillus bulgaricus*. He proposed that yoghurt consumption would prolong life and attributed this property to the organism.

Burnet, Sir Macfarlane A Nobel prize-winning Australian microbiologist. He demonstrated the value of hand-washing in recounting that during the visit of Queen Elizabeth II to Western Australia in 1954 while a poliomyelitis outbreak was underway, the compulsory washing of schoolchildrens' hands after using the toilet lead to a reduction in new cases.

Butchers' shops

- Risk of cross-contamination from raw to cooked meats is very high and foods and equipment must be kept separate.

- Food intended for sale as pet food must not be kept with human food.
- Adequate refrigeration must be provided.
- Disposal of waste can easily cause problems with pests and other animals such as dogs and cats.

Butter Due to its high milk-fat content (over 80%) butter is not known to have caused food poisoning. It is a perishable food and should be kept between 1–4 °C although it may be safely frozen for several months. Butter deteriorates due to:

1) the actions of micro-organisms which cause rancidity resulting in the production of glycerol and free fatty acids.
2) oxidation reactions forming organic peroxides. The taste and smell of rancid butter is both characteristic and objectionable and is due to the production of butyric acid.

Byelaws Local authorities may make legislation which only applies to a specified district or area. It must be approved by central government and can cover any aspect over which the local authority has control including food matters. One example is the banning of dogs entering food shops.

***Byssochlamys* moulds** These can affect some canned products. They are heat resistant, a comparitively rare property of moulds.

Cadelle beetle (*Tenebroides mauritanicus*) A distinctively shaped flat black insect about 15 mm long, the adult having a marked 'waist'. It feeds mainly on cereals and is often persistent in flour mills. The habit of larvae burrowing into wood provides harbourage for this and other pests.

Cadmium A metal used to electroplate steel and other metals such as those used in the manufacture of utensils and other equipment. Acids such as fruit acid may dissolve it causing chemical food poisoning. Cadmium-containing glazes used for pottery and some enamelware also suffer from this problem.

Calcium propionate A preservative used in dairy and bakery products, particularly against moulds and 'rope' organisms.

Campden tablets A disinfectant containing sodium or potassium metabisulphite often used in brewing for 'sterilising' equipment and finishing fermentations. The active agent is sulphur dioxide, a strong bactericide.

Campylobacter A group of Gram-negative bacteria causing food-borne disease. Food poisoning outbreaks caused by this group have shown a dramatic increase in numbers over recent years. In Britain in the early 1990s around 30 000 cases per year have been reported.

- The bacterium (a spirally-curved bacillus) is a non-spore former, with an optimum temperature of 43 °C.
- It is microaerophilic to anaerobic, and is sensitive to low pH.
- The onset period for the disease varies from two to eleven days.
- Symptoms, probably caused by the release of an enterotoxin, start by resembling influenza and include diarrhoea, which usually contains blood.
- The illness lasts from one day to several weeks with few fatalities.
- Complications such as meningitis and endocarditis (inflammation of the lining of the heart) are possible.
- Foods commonly involved are mainly raw poultry, but meat, offal, raw milk and contaminated water have also been found to be responsible for outbreaks.
- Contamination is usually from the excreta of humans and animals (domesticated, farm and wild, including birds).

In common with other food-borne diseases, small numbers of organisms are able to cause an infection. The usual species are *Campylobacter jejuni* and *coli*.

Candling A non-destructive method of visibly examining the insides of eggs by exposing them to a bright light. Various defects such as imminent hatching, large air spaces and blood spots may be seen.

Canning A method of food preservation developed in the early 1800s in France by Nicholas Appert who won a prize from the Emperor Napoleon for his efforts. It was originally called 'Appertisation'.

- The process relies on heat to destroy micro-organisms (pathogenic and spoilage), the aim being to achieve commercial sterility.
- The pH of food is important in determining the amount of heat required since pathogens (in particular *Clostridium botulinum*) generally do not grow below pH 4.5.
- Other factors affecting the length of heat treatment include the size of the can and whether or not the contents are liquid or solid; solid contents require a longer process.
- Cans are made fron tin-plated steel and may be lined with protective coatings or lacquers.
- The end-seam joining lid to body is important and must have the correct dimensions (overlap) to ensure proper closure.

- The headspace is normally a vacuum, this being measurable, and is an index of effective processing or shelf-life.
- Canned foods have shelf-lives varying from a few months (tomatoes) to many years (corned beef).
- Cans should be stored at 10–15 °C in dry conditions; food should not be stored in opened cans otherwise corrosion will occur.
- Canned meats may be stored in a refrigerator for a few hours before use to make slicing easier.

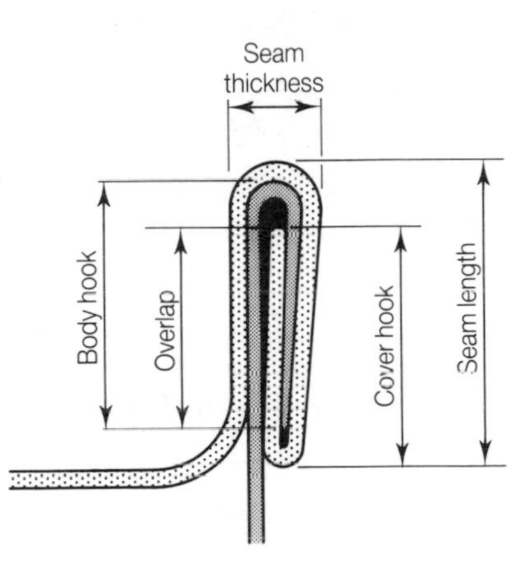

Can end-seam

- Long term storage of cans in refrigerators may lead to corrosion (rust) and then contamination.
- Canned foods are generally very safe but any high-risk canned food must be treated as though it were fresh once the can has been opened. For example opened canned ham must be refrigerated (in a suitable container) and used within a few days.

See: Aberdeen typhoid outbreak, Acid foods, *Bacillus stearothermophilis*, Blown cans, Botulinum cook, Fo, Sterilisation, Sulphiding.

Capon Male chicken killed at 12 weeks. It is sometimes treated with a hormone implant to increase speed of body weight production.

Capsule The bacterial capsule (also known as a slime layer) is a complex structure and forms the outermost layer of some types of bacteria.

Its presence has two main effects:

1) to confer virulence on pathogenic bacteria, and

2) to cement together those forms that exist in groups or aggregates of cells, e.g. the *Staphylococci*.

The specific composition of the capsule is the basis of serotyping of bacteria, which otherwise are identical. *See*: Serological typing.

Carbon dioxide A gas used to control growth of spoilage organisms such as the aerobes *Pseudomonas* and *Achromobacter*. Examples include storage of chilled meats, fruits and vegetables at an atmosphere of up to 10% carbon dioxide. The gas may be added in large storage rooms or be the result of wrapping food in selectively-permeable films.

Carcases All food animals must be inspected by a qualified person after slaughter. *See*: Appendix 1: Food Law, BSE.

Carcinogenic The ability of an agent to cause cancer; such substances include many chemicals and various forms of radiation.

Carpet beetle (*Anthrenus* species) A wide variety of small beetles (up to 4 mm long) whose body scales give them a variegated appearance (patches of different colours). They occasionally infest food but are more commonly found in woollen materials and fabrics.

Carrier An individual who harbours disease-causing organisms inside the body and excretes them, without suffering from or showing any symptoms of the particular disease.

Incubation carriers
Most people excrete organisms during the incubation period of a disease without showing symptoms.

Convalescent carriers
People who excrete organisms while recovering from a disease, with symptoms having ceased.

Healthy carriers

- This is the most potentially dangerous group to public health particularly if they handle or prepare food.
- These food handlers show no signs of illness at any time, but they excrete the pathogenic organisms.
- They are unlikely to pay the necessary extra attention to their personal hygiene, unlike incubation and convalescent carriers.

Carriers may also be described as:

1) chronic carriers, who excrete the pathogenic organisms throughout their lives;
2) casual carriers, who excrete the pathogenic organisms for a short period only, e.g. weeks or months.

All carriers are likely to spread disease. They are difficult to identify and therefore treat, both of which are required in order to stop them from excreting the organisms. Carriers are also known by the medical term 'symptomless excreters'.

Carveries These have a particular food hygiene hazard in that cooked meats are offered for sale for lengthy periods. They may also be subject to lack of temperature control and problems of contamination. Overhead infra-red heaters are sometimes used to maintain temperatures. The maximum time cooked meat may be kept (displayed) at below 63 °C is four hours, and no more should be displayed than is reasonably necessary.

Cash (handling of) There is evidence that money, both coins and notes, can provide a temporary residence for micro-organisms which could then be passed on to food. Control measures could include hand-washing or the use of separate staff, although practically the risk is limited.

Casual staff Persons who occasionally help out may not be aware of the problems of hygienic food handling and care should be taken that their knowledge of this is sufficient. Training of such staff should be such that food is safely produced and their work practices should be of the same standard as other food handlers. *See*: Due diligence.

Cationic detergent A surface-active molecule with a positively-charged active group (ion). Often slightly acidic (pH 6–7), they have limited powers of detergency but are bacteriostatic and find uses as disinfectants in sanitisers. An example is the variety of quaternary ammonium compounds.

Cats Live animals carry many diseases which may be transmitted to humans by way of food and so cats should not be allowed in food rooms or to use utensils etc used by people. Hairs and other items may also contaminate food. *See*: Animals

Caustic potash An alkaline cleaning agent possessing corrosive powers similar to caustic soda.

CDR (Communicable Disease Report) A weekly bulletin, produced by the Central Public Health Laboratory (CPHL), circulated to EHOs and the medical profession, giving details of food poisoning and other communicable diseases.

CDSC (Communicable Disease Surveillance Centre) Part of the Central Public Health Laboratory (CPHL) dealing with food poisoning and other diseases.

Ceilings All parts of food rooms including ceilings should be constructed so that they can be maintained in a clean condition, even though they are not surfaces which normally come into direct contact with food. A gloss painted surface would be adequate, as long as flaking did not occur. The time between cleaning and redecoration is judged according to needs. *See*: Design of food premises.

Cell membrane A structural component of all cells. In animal cells it encloses the contents of the cell, but in the bacteria (and plants) the cell membrane is surrounded externally by a tough cell wall. The cell membrane is selectively permeable, controlling the movement of substances into and out of the cell. In the bacteria it is concerned with cell respiration and, in some pathogenic forms, the production of toxins. The outer membrane contains molecules which are toxic to humans. *See*: Endotoxins.

Cell wall (bacterial) Encloses the bacterial cell. It has a fairly rigid structure and gives shape to the bacterial cell. There is a procedure, known as the Gram stain, which assigns all bacteria to one or other of only two groups, depending on the result of the stain. It is a useful aid to identification. The different reactions to the Gram stain are thought to be due to differences in the composition of the cell wall.

Central Public Health Laboratory (CPHL) Located at Colindale (London, England), this is a a major research institute concerned with food microbiology. *See*: PHLS.

Centre temperature (Core temperature) The centre of food is the last part to reach the desired temperature during cooking or cooling and is often measured using a probe thermometer. Key temperatures are:

- 70 °C should be reached during cooking and reheating
- 63 °C is the legal minimum for hot food
- −18 °C for frozen foods
- 1–4 °C for food in refrigerators.

Important points

- Air temperature in equipment (e.g. in an oven or refrigerator) may be different from the temperature of the food.
- The outer temperature of food will be different from the internal temperature.
- Probes must be positioned accurately.
- Probes must be disinfected after use.
- Temperatures may be taken at several points.

See: Appendix 1: Food Law.

Cereals Not often associated with food poisoning except for outbreaks involving the bacterium *Bacillus cereus*. Storage of cereals and related products must be such that pest infestations, including beetles, other insects and rodents are prevented. In damp conditions, mould growth is a problem. *See*: *Aspergillus*, Dry goods store, Rice.

Cestoda (Cestodes) *See*: Tapeworm species.

Changing facilities A separate room in which staff may change from outer clothing into workwear and in which outer clothing and footwear may be stored is desirable. It is illegal in Britain to keep outer clothing in food rooms unless it is in a cupboard or locker. This is to prevent contamination of food by dust, clothing fibres, etc.

Chadwick, Edwin One of the founding figures of public health law in England during the middle of the nineteenth century and the first president of the Sanitary Inspectors' Association (1883). The current headquarters of the Institution of Environmental Health Officers is in Chadwick House, London.

Cheese Milk products made in a variety of ways from assorted animal milks by the action of bacteria such as *Streptococcus lactis*.

- Hard cheeses of the cheddar type are not high-risk foods.
- Cheeses ripened with moulds or micro-organisms must be refrigerated.

Outbreaks of food poisoning from cheeses have been due to *E. coli*, *Listeria monocytogenes*, *Salmonella* species, *Staphylococcus aureus* and others. Cheeses made from raw milk are liable to contain many species of micro-organisms including pathogenic moulds. *See*: Appendix 1: Food Law, Milk.

Cheese mite (*Tyrophagus casei*) A pest of cheese and other foods which is also responsible for the characteristic flavour and surface appearance of Altenburg cheese. When ripe, the cheese is covered with a grey-white powder consisting of live and dead mites.

Chemical disinfectants/sterilants *See*: Hypochlorite.

Chemical food poisoning Many 'one off' cases of food poisoning and other diseases have been noted due to accidental contamination of foods with chemicals ranging from pesticides to cleaning chemicals. Two infamous events include:

1) mercury contamination of fish in Minamata Bay, Japan, in the 1950s which caused congenital deformities and brain damage

2) an incident in Spain in the 1980s in which hundreds of people died as a result of consuming contaminated cooking oil.

Chickens

- On a live chicken, and other birds, there are vast numbers of bacteria on the skin, and others from the feathers and faeces.
- Contamination may come from feedstuffs, such as fish meal and animal protein. Pasteurisation of such feedstuffs is recommended.
- Contamination of the skin and the lining of the body cavity occurs at various stages in and following slaughter, such as washing, plucking and evisceration (removal of innards).
- There is increasing evidence of contamination of eggs with *Salmonella* when laid.
- The giblets (gizzards, hearts and livers) are also frequently contaminated.
- In the 1970s and 1980s, 60 to 80% of chicken carcases were found to be contaminated with species of *Salmonella*.
- During the early 1990s less than 50% *Salmonella*-contamination was reported. (*Note*: Figures quoted refer to Britain.)

The following bacteria have been isolated from poultry and poultry products: *Alcaligenes, Bacillus, Corynebacterium, Enterobacter, Escherichia, Flavobacter, Proteus, Pseudomonas, Sarcina (Micrococcus), Salmonella* and *Staphylococci*; in addition a number of yeasts have also been isolated.

Consequently, careful thawing, storage, and thorough cooking of all chickens (indeed all poultry) are essential. It is also important that any blood from poultry is not allowed to contaminate other food. *See*: Cross-contamination, Eggs (various entries), *Salmonella* (various entries), Thawing of food.

Chilled foods Foods kept under refrigeration, normally between 1–4 °C, to extend their shelf-life by a few days. Such foods are perishable and often high-risk. Temperature control is required by law in the UK. *See*: Appendix 1: Food Law, Cook-chill, Danger zone, Refrigerators.

Chlorine

- The hypochlorites (chemical disinfectants) release chlorine, which is toxic to many micro-organisms due to its powerful oxidation of structural protein, causing death of cells.
- This action takes a few minutes and sufficient time (contact time) should be allowed for when an item is to be disinfected.
- Effectiveness is hindered by heavy soiling of surfaces by food residues.

- They also have a strong taste and odour, must be rinsed from food contact items and may corrode metals.
- The non-biodegradability of hypochlorites is also an issue, but these substances are both cheap and effective.

See: Disinfectants, Hydrogen peroxide.

Chocolate Confectionery products high in both fat and sugar, rarely involved in food poisoning. In 1982–3 an outbreak of food poisoning due to the bacterium *Salmonella napoli* occurred across Europe under very uncommon circumstances, including a very low infective dose. A more frequent cause of complaint with chocolate is the presence of a white surface deposit due to migration of some ingredients over a long storage period. Generally chocolate is a low risk food which may be safely stored at room temperature.

Cholera

- A disease of the gastro-intestinal tract caused by the bacterium *Vibrio cholerae*.
- It is transmitted via the faeces through water, contaminated foods (particularly shellfish harvested from polluted water) and rarely from cholera carriers.
- The incubation period is around two to three days.
- The effects on the sufferer are due to an enterotoxin which causes diarrhoea, which may be as much as 20 litres (35 pints) per day; there is also violent vomiting.
- The severe fluid loss, without treatment, results in death.
- Treatment requires replacement of lost body fluids, together with antibiotics.

It is a rare disease in Britain, due to efficient sewage disposal and water purification.

Chopping boards

Wooden boards

- Traditionally used, but wood is increasingly considered to be an unsuitable material.
- Wood is difficult to clean, it is easily cut and it absorbs moisture.
- Wooden chopping boards must be in excellent condition if they are to be used for food preparation.
- The practice of salting boards after use does not have any great effect on *Staphlyococci*.
- Wooden boards may harbour bacteria from some foods prepared on them and so contaminate other foods later prepared on the same board.

Polypropylene boards

- Modern materials, such as polypropylene, are now available and are claimed to be tougher and non-absorbant.
- They may be colour coded, for example red for raw meat, green for vegetables, blue for fish.
- These boards, in common with wood, may be deeply scored and become difficult to clean.
- Some users report a safety problem in that plastic boards are slippery.

Ideally separate blocks and boards should be used for different foods, particularly for raw and cooked foods. This helps to control cross-contamination.

Chromobacterium lividum A Gram-negative spoilage bacterium causing greenish-blue to brownish-black spots on stored beef.

Chronic disease A disease that develops slowly and is usually not very severe, but is continuous or recurrent for long periods of time.

Cigarette beetle (*Lasioderma serricorne*) A pest of tropical products including cocoa, mainly infesting tobacco during storage and manufacture. The adult resembles the biscuit beetle but is smaller.

Ciguatera A type of neurotoxin food poisoning similar to paralytic shellfish poisoning, caused by a number of tropical species of fish including Caribbean varieties. It causes at least 50 000 cases per year world-wide but is very uncommon in the UK. The causative agents are marine algae (dinoflagellates) including *Gambierdiscus toxicus, Prorocentrum concavum* and *P. mexicana*. The disease may be fatal (up to 20% mortality).

CIP (Cleaning-in-place) systems A method used to clean equipment, often involving pipework, without first dismantling it. Cleaning chemicals and rinses may be pumped through equipment to remove contamination, food residues and deposits. Simple systems are found in the licensed trade and are used to clean beer pipes, while more complex operations exist in food factories, including milk processing plants.

Cladosporium A group of green moulds which are responsible for spoilage of a wide range of foods, including fresh meat, fruits, vegetables and cheese. They can grow at temperatures down to – 4°C. The most frequent spoilage agent is *Cladosporium herbarum*, the growth of which appears dark green to black. The spores of this organism are very commonly found in the atmosphere. Other species cause green, brown or black discolourations. Examples of spoilage include:

1) black spot on meat, especially if freshly butchered and stored at 0 to 4°C, and particularly if the surface of the meat dries out;

2) green mould rot of stone fruits, e.g. peaches, apricots, plums and cherries;
3) cheese, where the mycelium and spores are dark or smoky in colour, producing dark colours on the cheese;
4) another member of this group is *Cladosporium cladosporioides*, which causes softening of cucumbers.

Clean-as-you-go A motto in food hygiene indicating that it is the food handlers' personal duty to clean up food waste caused by themselves thus avoiding any build-up. Fresh deposits are easier to remove than older ones.

Cleaning and cleaning schedules Often described as being the largest day-to-day maintenance operation in any food business.

Problems

- Cleaning is often regarded as a low status job and is frequently neglected.
- Cleaning staff are often untrained.
- Inadequate cleaning and the subsequent build up of deposits is high on the list of reasons for immediate closure of food premises and prosecutions.
- A food business may be judged on its cleanliness by both customers and enforcement officers.

Solutions

- In the ideal situation, cleaning should be performed by a specialist team who arrive fresh on the job with the means to operate effectively. In practice, cleaning is frequently performed in a haphazard way by tired staff who have just finished a long shift of food production.

Cleaning schedules

The key to managing cleaning operations is to systemise the approach by preparing a schedule stating:

WHAT?	The item to be cleaned.
HOW?	The method to be used, including use of chemicals, equipment needed and safety precautions.
WHEN?	How often? For example, daily.
WHO?	The person responsible.
MONITORING	Check that cleaning has been achieved, for example, visual or microbiological.

Cleaning schedule

The schedule, or plan, should include details of all equipment needed and state any safety precautions necessary. It must be capable of being checked or monitored to ensure that the system is working. Checking may be visual or microbiological (by taking swabs). Complete cleaning schedules may be purchased 'off the shelf' from cleaning chemical suppliers.

Cleaning-in-place systems *See*: CIP

Cling film A thin plastic self-sticking film used to cover foods both during storage and in certain cooking methods including microwaving.

- It is important to use the correct grade of film. The plasticiser chemical used in some film has been reported to be both fat-soluble and carcinogenic and its use in contact with fatty foods is not advised.
- The space between food and cling film may have a very high humidity and provide favourable conditions for mould growth if storage is long-term.

Cloakrooms Food handlers must have somewhere away from the food rooms to place their non-work clothes. This may take the form of a changing room or something similar. If clothes are taken into food rooms then lockers or cupboards, must be supplied. *See*: Appendix 1: Food Law.

Clostridium botulinum An important and deadly food poisoning bacterium.

- A Gram-positive anaerobic bacillus and a spore-former, so the organism will survive high temperatures.
- Its ability to spore makes it the most heat resistant of all food poisoning organisms.
- Causes a disease known as botulism, the food poisoning being due to the release of an exotoxin.
- Toxin production is least likely when the environment is aerobic, below 5 °C, below pH 4.5, and in the presence of nitrites, as in smoked meats and fish.
- The onset period for the disease is usually 18 to 36 hours (exceptionally 12 to 96 hours).
- Duration of the illness varies from death in 24 hours to eight days, or slow convalescence over six to eight months.
- First symptoms include slurring of speech, double vision and loss of balance since the toxin affects the nervous system (it is a neurotoxin).
- Death may result due to paralysis of various brain centres such as that involved in the control of breathing.

Foods commonly involved are any food which is not acid (above pH 4.5) particularly in any canned or bottled container, which provides anaerobic conditions. Contamination of soil, meat and fish is likely since spores of the organism are widely distributed in nature.

Clostridium difficile A Gram-positive bacterium which has been found to be responsible for a number of outbreaks of diarrhoea in London and Staffordshire (England) in the early 1990s. Affected individuals were hospital in-patients in geriatric wards. It is thought that the organisms were allowed to grow in the bowel due to the effects of antibiotic therapy, which had been prescribed for other medical conditions. Some strains of the organism produce a weak toxin and have an optimum temperature for growth of 30 to 37 °C.

Clostridium nigrificans *See*: *Desulphotomaculum nigrificans*.

Clostridium perfringens

- A Gram-positive anaerobic bacterium and a spore-former, so often survives normal cooking.
- Grows at between 15 to 50 °C.
- Causes food poisoning as a result of change to endotoxin released in the digestive system.
- The illness is often due to the organisms multiplying during long, slow cooking and storage.
- It usually requires extremely large numbers of organisms to be present in the food for the illness, which is relatively mild, to occur.
- The onset period for the disease is 8 to 22 hours and the duration of illness is 12 to 48 hours.
- Symptoms are mild, with abdominal pain and diarrhoea, rarely accompanied by vomiting, but often with nausea.

Foods commonly involved are meat dishes, such as stews or rolled joints, particularly when they are not cooled rapidly and refrigerated after cooking, or if they are not re-heated thoroughly.

Contamination is usually from excreta, raw meat, poultry and soil. *Note*: This organism was formerly known as *Clostridium welchii*.

Clostridium sporogenes A Gram-positive spoilage bacterium of canned foods. It is a proteolytic organism (i.e. it breaks down proteins), producing foul smelling (putrid) compounds. The food may be partly digested. Affected cans swell and may even burst. It affects particularly canned meats and fish. Spoilage is ordinarily prevented by adequate refrigeration, and the cause is due to underprocessing.

Clostridium thermosaccharolyticum A Gram-positive spoilage bacterium of canned food. It is a saccharolytic organism (i.e. sugar-splitting), and causes gaseous spoilage of canned vegetables, producing a sour, fermented, cheesy odour. It typically forms acid and gas in low and medium acid foods, and affected cans swell and may burst.

Clostridium welchii The former name for the bacterium *Clostridium perfringens*.

Closure orders *See*: Emergency prohibition notice and order, Improvement notice.

Clothes moth (*Tineola bisselliella*) Occasional pests of stored products. *See*: Moths.

Clothing

- Protective clothing is worn to protect food from risk of contamination and not to keep other clothes clean.
- Persons handling open food must wear clean and washable over-clothing at all times.
- Light coloured uniforms are preferred (ensures cleanliness and soiling is obvious).
- The design of such clothing should minimise physical contamination of food by having no pockets or buttons.
- Comfort of the wearer should not be neglected; some fabrics are better than others in this respect. Nylon, for example, may cause sweating.
- Colour-coded uniforms may be appropriate in operations where the prevention of cross-contamination is vital.

Other items of clothing may include head and beard coverings, gloves, footwear and wellington boots. All of these must be capable of being cleaned and kept in that condition.

Cloths These are often implicated as the vehicle by which food poisoning bacteria are spread from place to place. Cloths must be used with great care, the disposable type being preferred in many situations. Cloths for re-use should be disinfected with hot water or chemicals.

Cobalt 60 A radioactive isotope finding industrial and medical use as a source of X-rays. *See:* Irradiation.

Coccus A term which describes the shape of certain bacterial cells; such cells are spherical. *See*: Bacterial classification.

Coccidiosis A protozoan infection in rabbits caused by *C. serialis* and *C. pisiformis* resulting in small white cysts in the abdomen and yellowish areas in the lungs and liver. Affected animals are condemned.

Cockroaches

- Insects causing widespread infestation of food stores and related areas, particularly those which are heated.
- Cause physical contamination.

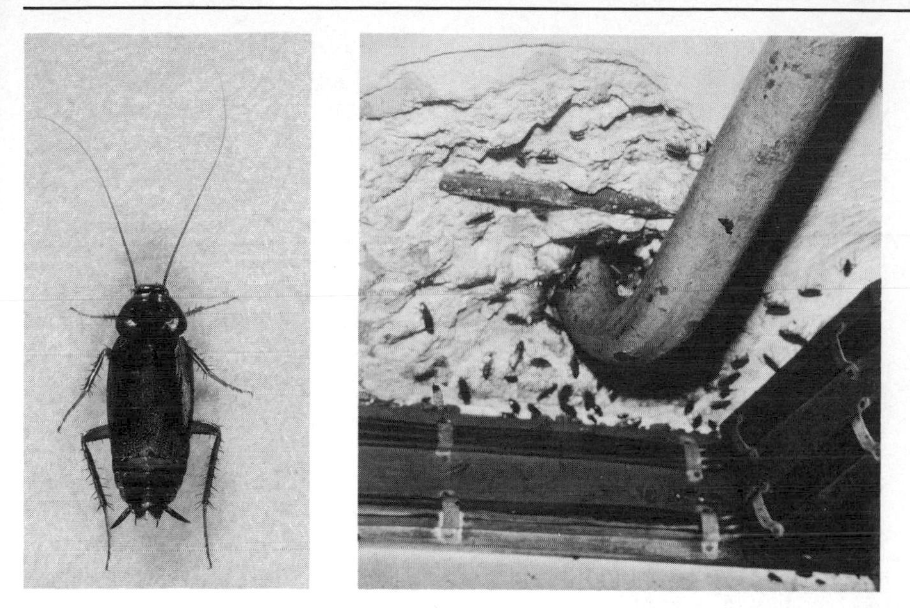

Cockroaches

- They rarely exist in small groups and are generally nocturnal.
- Infestations may be very difficult to deal with. Eggs are laid in numbers between 16 and 32 in protective egg cases called ootheca which enable them to survive in poor conditions for long periods even though the adults may be destroyed by a programme of treatment.
- As with other pests, treatment is a professional task involving insecticides and poison baits. Juvenile hormone may also be used.
- Several species are known, including:

  1) the American cockroach (*Periplaneta americana*), reddish-brown in colour, up to 40 mm long
  2) the German cockroach (*Blattella germanica*), pale brown in colour, up to 15 mm long
  3) the Oriental cockroach (*Blatta orientalis*), dark coloured, up to 25 mm long.

All are known potential carriers of food poisoning organisms such as *Salmonella*.

Cocoa moth (*Ephestia elutella*) Also known as the warehouse or mill moth, it infests chocolate and tobacco as well as general foods.

Coconut *See*: Desiccated coconut.

Codes (date) Date-coding of goods gives information to suppliers and consumers regarding the shelf-life of products and is also useful to investigators if problems arise. Many systems are in operation from 'secret' coding to open dating. *See*: 'Best-before' date, Stock rotation, 'Use-by' date.

Codes of Practice Advisory or guidance documents produced by various bodies including government and professional organisations. As such, these have no legal weight but if a Code of Practice is ignored and a statute or other law is broken, then this would be evidence that insufficient care had been taken. There are also statutory Codes of Practice which give guidance to environmental health officers, amongst others, on how to enforce certain legal provisions in a uniform way.

Codworm (*Phocanema*) A brownish-white round worm, up to 3 cm long, found around the liver and intestines of fish, and sometimes entering the flesh. Freezing and curing kills them but curing alone often does not. There is no evidence that the worm affects humans but badly infested fish are aesthetically unfit.

Cold chain A term used to describe the unbroken storage requirement of chilled high-risk foods from production to consumption. Such a chain includes all storage and transport used by all persons whether manufacturer, wholesaler, retailer or consumer. *See*: Temperature control.

Cold (effect on bacteria) *See*: Growth requirements (bacterial).

Cold spots (in microwave cooking) Microwaves may be deflected (refracted) by ice crystals in frozen food so that some areas receive little heating and consequently remain frozen. Rotating or moving the food during heating helps to overcome the problem. *See*: Microwave ovens, Runaway heating.

Cold storage *See*: Chilled foods, Freezers, Refrigerators.

Cold store taint The absorption of odours by foods during cold storage. The use of containers with lids and proper wrapping reduces this risk.

Coliforms

- A group of bacteria which are commonly found in the large intestine (or colon, hence the name).
- They make up a large part of the intestinal flora of man and other vertebrate animals where they are normally kept under control.

- Examples of such organisms include *Enterobacter aerogenes*, *Escherichia coli*, *Shigella dysenteriae*, and *Vibrio cholerae*.

  Note: These organisms are strictly referred to as **faecal** coliforms, since other coliforms are more commonly found in plant and soil samples. Their presence is generally undesirable in foods, since if they are detected in some foods, e.g. oysters, it is considered to be an indication of sewage contamination and so there is the possibility of intestinal pathogens being present. Also their growth in foods results in spoilage, causing the production of 'off' flavours.

Colon The large intestine.

Colony (bacterial) A population of micro-organisms which have grown on the surface of solid medium (commonly agar) as the result of the reproduction of one or a few cells. Colonies have characteristics, such as texture, size, shape and colour, and these are fairly constant for a given species. Such characteristics are useful in distinguishing one micro-organism from another with the naked eye.

Colony count Microbiological technique involving counting the number of colonies of organisms after a specified time period. Uses of such figures include quality control of raw materials and products and assessment of cleaning processes.

Colour coding A system of controlling cross-contamination by using different coloured equipment for individual foodstuffs or in separate areas. It is common to have red as the raw meat colour code. Typical coded items could include chopping boards and knife handles. *See*: Clothing.

Commensalism A form of symbiosis in which one individual benefits but the other is not harmed. Such relationships exist in many forms of life, including man and a variety of micro-organisms, such as certain bacteria and yeasts. Examples include *Streptococcus viridans* (a bacterium) found in the mouth, and *Candida albicans* (a yeast) found on the surface of the skin.
Note: Such a relationship may change under certain conditions and the organisms may become pathogenic.

Commercial sterility The near-total destruction of all micro-organisms and spores such as is achieved in canning. Some non-pathogens will survive but the product is judged to be safe. *See*: Sterilisation.

Communicable disease A disease capable of being transmitted from one person to another, e.g. food poisoning. *See*: Notifiable disease.

Compaction Large volumes of waste packaging materials such as cardboard boxes may be reduced in size by pressing them into bales in a compactor. Such bales have an additional advantage in that harbourage for pests is reduced.

Compensation In the event of an enforcement officer such as an EHO making a professional mistake which is proved to be harmful to a business or person, compensation may be claimed. An example would be for losses suffered in complying with an emergency prohibition notice served on a business, unless an application is made for an emergency prohibition order within three days *and* the court is satisfied that the health risk condition was fulfilled.

Condensed milk

1) The sweetened type is a high-sugar product, not considered to be high-risk. Preservation is achieved by a combination of osmosis and heat.
2) The unsweetened type undergoes a severe heat treatment and is as safe as other canned foods. If water is added after the can is opened then this product must be treated in the same way as fresh milk.

Conduction A method of heat transfer in solid materials in which molecular vibration transfers heat to adjacent molecules. Cooking methods using conduction including griddling and the heat from microwaves travels in solid foods by this method.

Confectionery Products with a high sugar content, not considered to be high-risk foods.

Confused flour beetle (*Tribolium confusum*) A dark coloured insect, around 6 mm long, being a pest of stored cereal products. *See*: Beetles.

Conidiophore A fruiting aerial hypha, produced by some mature fungi, bearing conidiospores.

Conidiospore A type of reproductive asexual spore produced by some types of fungi. It may be unicellular or multicellular, and is not enclosed in a sac. Such spores are produced in chains at the end of a conidiophore. *Penicillium* (a common spoilage mould) reproduces by means of conidiospores.

Consumer protection A large amount of legislation intended to protect consumers is enforced by various agencies, including Trading Standards Officers. Some food issues are included, such as the weight of prepacked goods.

Contact dust Rodenticidal or insecticidal powder not for food room use. Poison may be eaten by the animal during grooming or may be taken back to nests.

Contact time The length of time taken for a disinfectant chemical to achieve the desired effect. It is normally in excess of three minutes for hypochlorite on food-contact surfaces.

Contamination of foods Most natural foodstuffs are either of plant or animal origin, and both have their particular surface populations of micro-organisms. Animals also have an intestinal population and they also liberate micro-organisms in their excretions and secretions. Both groups may, from time to time, suffer from parasitic disease, and so they will also carry those particular pathogens.

Contamination may also occur from an outside source; the main sources being:

- green plants and fruits (bacteria, yeasts and moulds) animals, including meat, hides, hooves, hair, faeces, and for poultry, feet, feathers and eggs (mostly bacteria)
- sewage (bacteria and viruses)
- soil (bacteria, yeasts and moulds)
- water (bacteria and viruses)
- air (bacteria, fungal spores and viruses)
- food handlers (bacteria and viruses)
- kitchen surfaces and equipment (bacteria, yeasts and moulds).

Controlled atmosphere In the storage of meats, fruits and vegetables the addition of carbon dioxide is used to slow microbial growth and delay ripening. In some fruits such a pears and melons the addition of the gas ethylene hastens ripening at the end of the desired storage time.

Convalescent carrier *See*: Carrier.

Convection A method of heat transfer occuring in gases and liquids by convection currents. Since heat rises, these currents move away from the heat source and circulate throughout the substance being heated. In the reheating of large volumes of food such as soups and stews it is important that convection does not leave cool spots.

Cook-Chill A method of interrupted catering in which food is

1) prepared
2) cooked to a core temperature of 70 °C
3) rapidly chilled to 3 °C
4) stored at 3 °C (this may include transport)
5) regenerated (reheated) to 70 °C
6) eaten immediately.

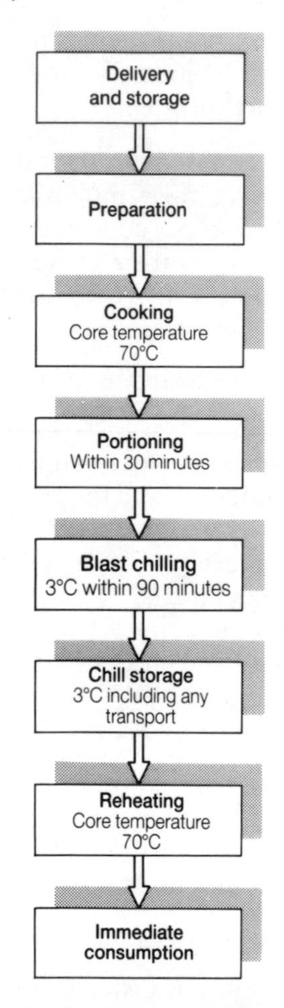

Cook-Chill procedures

The whole process must not take more than five days. In practice this means a day for production, three days storage and a day for reheating and consumption. There is a set of guidelines, produced in 1989, available from the Department of Health regarding cook-chill procedures entitled, *Guidelines on Cook-Chill and Cook-Freeze Catering Systems*.

Temperature control at all stages of the process is vital and probe thermometers are frequently used. If food during storage reaches 5 °C it must be eaten within 12 hours. If 10 °C is exceeded the food must not be eaten.

The system, which operates on a factory basis, must be carefully controlled at all times since there is a danger from *Listeria monocytogenes*. Typical cook–chill users include hospitals and supermarkets with the end-users (consumers) often some distance away from the place of production. Temperature control must be maintained during delivery.

Advantages of cook-chill

- Wide menu choice and increased palatability. 24 hour production.
- Economies in purchasing raw materials in bulk.
- Centralised standards.
- Better portion control.

Cook-Freeze A system similar to cook-chill but with an extended shelf-life due to storage at −18 °C.

Cooking of food (effect on bacteria)

- Fresh food, cooked thoroughly and served immediately, should never cause food poisoning due to micro-organisms.
- If food is cooked thoroughly it should ensure that all bacteria in and on the food will be killed (except for spore formers).
- The safest method of cooking is to use a short, high-temperature method; this means that the food is quickly taken through the 'danger zone' of bacterial growth, to reach a temperature at which the death of bacterial cells starts to occur.
- To ensure that bacteria are destroyed, cooking times and temperatures must be carefully calculated.
- It is important to realise that cooked food may cause problems if it is allowed to cool too slowly, or if it is allowed to be re-contaminated.

See: Cooling of food, Cross-contamination.

Cooking oils In use, these reach temperatures in excess of 160 °C. Care is needed, particularly with the cooking of frozen foods, since it may result in the surface of a product being cooked but the centre temperature (core temperature) being inadequate for thorough cooking.

Cooling of food

- Cooked food required for cold service after cooking must be cooled to a temperature of between 10–15 °C **within 90 minutes** to avoid the multiplication of bacteria which have survived the cooking process, or the germination of spores which have not been destroyed.
- Large single items of food such as meat may need to be cooked in pieces no heavier than 2.5 kg (6 lb) while large volumes of liquid (more than 25 litres, or 44 pints) may be split into smaller shallow portions to help rapid cooling.
- Hot food must not be placed directly in a refrigerator.
- Cooling may be done in a cool room (10–15 °C) or in a blast chiller.

Cooling water In the canning industry, contaminated water has been a cause of food poisoning. *See*: Aberdeen typhoid outbreak.

Copper One of a number of metals which may cause food poisoning. It can enter food from water pipes or from the action of acid foods on copper pans. *See*: Metals.

Corynebacteria A group of Gram-positive bacteria, some members of which are pathogenic to humans and animals. Some cause disease and spoilage in plants, such as tuber rot of potatoes and fruit spot of tomatoes.

Corned beef Canned meat product ('corn' originally meaning salted) which although generally safe has been involved in food poisoning. *See*: Aberdeen typhoid outbreak, Canning.

Cornflour A potential source of *Bacillus cereus* organisms.

COSHH (Control of Substances Hazardous to Health) Regs 1988 Legislation requiring employers to assess the risks in the use of all chemicals which may be a danger to health. These are defined under separate regulations and include toxic, harmful and corrosive substances. Employees must receive training and information regarding uses and risks. Those handling very dangerous chemicals may require regular health checks. In the food industry the major COSHH problems are found in cleaning chemicals such as bleaches and caustic soda. *See*: Cleaning (various entries).

Coughing (and open food) Coughing results in a violent release of micro-organisms from the mouth and throat into the air. Many of these micro-organisms are in small clusters, surrounded by mucus. This helps to prevent them from drying and so they remain infective for longer. Any such clusters of bacteria landing on unprotected high-risk food may well be the starting point of an outbreak of food poisoning. *See*: *Staphylococcus aureus*.

Covering of food In order to prevent airborne contamination, open food (for example that on display in shops) must be covered. *See*: Aerosols, Dust.

Courts Most food hygiene offences will be heard at a local Magistrates Court (Summary offences) and there is provision for appeals from these to go to a Crown Court. At higher levels are the Court of Appeal, House of Lords and European Courts.

Cream A high-risk milk product, often pasteurised or sterilised, which must be kept refrigerated. Various legal standards define the composition of creams (e.g. double cream is 48% fat). Hygiene precautions for cream also extend to imitation cream. *See*: Danger zone, High-risk food.

Crockery Chipped or cracked crockery should be discarded since the damaged areas are porous and may harbour bacteria.

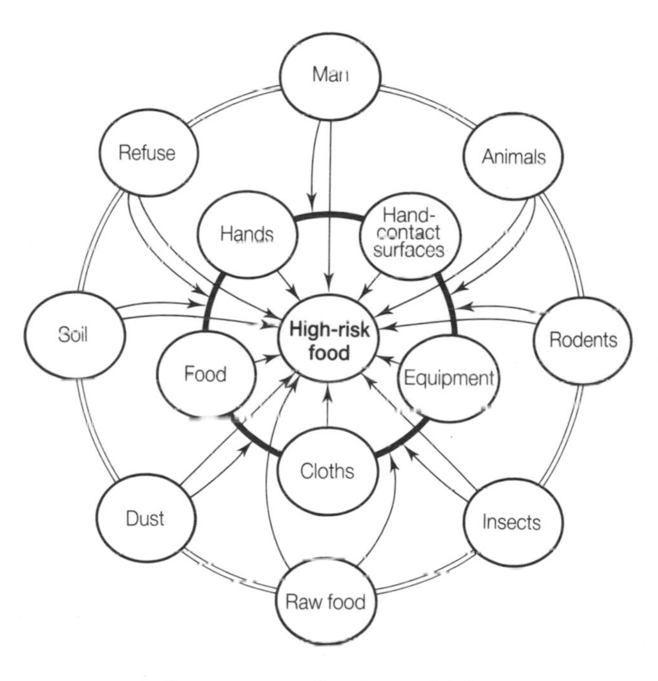

Cross-contamination vehicles

Cross-contamination The transfer of bacteria from one food, often raw, to another food, often cooked. Cross-contamination requires a vehicle. Examples are:

- using the same knife to cut raw and cooked meats, or handling them without washing the hands in between
- wiping up meat juices from a raw chicken with a cloth, and using the same cloth to dry a plate used for sandwiches
- failure to wash the hands properly after using the lavatory and then preparing sandwiches or a trifle
- preparing salad vegetables on a chopping board, the soil from which may contaminate a hard-boiled egg later sliced on the same board
- washing the hands thoroughly but then drying them on a soiled cloth prior to preparing food.

Crown immunity Premises occupied by Crown agencies may be excluded from the provisions of various laws. As an example, prior to 1 February 1987, food hygiene law did not apply to hospital kitchens. Some Crown premises are still exempt from both food hygiene regulations and health and safety legislation.

Crustacea A term used to describe a class of arthropods (literally 'jointed limbs') including crab, crayfish, lobster, prawn, scampi, shrimp and woodlice.

Cryogenic freezing A type of quick freezing in which food is immersed in liquid nitrogen. The process is extremely rapid but expensive and tends to be used for high-value products such as strawberries and prawns.

Cryovac (W R Grace Ltd™) The vacuum packing of fresh meat in a special plastic laminate which prevents the entry of oxygen and moisture and also physically protects the product. Shelf-life is said to be doubled at chill temperatures. *See*: Anaerobic, *Clostridium botulinum*, Sous vide, Vacuum packing.

Cryptolestes Small beetles (2–3 mm in length) sometimes infesting damp stored foodstuffs. The red-rust flour beetle is one example.

Cryptosporidium A unicellular parasite belonging to a group of micro-organisms known as coccidial protozoa. It was considered to be a commensal up to the mid 1950s when it was found to be a cause of fatal enteritis in turkeys. It was later found to cause diarrhoea in a wide range of animals, and in 1976, was shown to be able to cause illness in humans. The most common species responsible for infections in humans and livestock is *Cryptosporidium parvum*. The organism can only grow in living hosts.

- Infection occurs when cysts (known as oocysts) are swallowed and following a complex life cycle, faeces containing more infectious cysts are released from the body.
- The cysts are highly resistant to disinfectants, including bleach.
- They are killed by freezing (−20 °C for 30 minutes), by heat (55 °C for 5 to 10 minutes, or a few seconds in boiling water), and by some chemicals.
- The disease, known as cryptosporidiosis, most commonly affects children aged one to five years.
- The incubation period is about seven days.
- Symptoms include watery diarrhoea, abdominal pain, vomiting and fever, and last from one to three weeks.
- Source of infection may be animal or human and person to person spread is not uncommon.
- Water, raw milk and offal have all been identified as vehicles of transmission.

Culture media There are many media available for supporting the growth of micro-organisms in the laboratory. Essentially there are two types of media: liquid media (for bacteriological work) known as broths, and solid media, commonly nutrient agar gel. It is possible to select for the growth of particular micro-organisms by including in the broth or agar gel something to actively encourage the growth of those organisms. Conversely, substances can be used to discourage or inhibit the growth of all micro-organisms except for the one desired. There are special solid media for culturing fungi, again based on agar, such as potato dextrose agar (PDA) and malt agar, containing crushed boiled potato and malt extract respectively. Both bacteria and fungi will show appreciable growth in or on such culture media within 24 hours or so in the laboratory. Such laboratory isolation and growth is a vital first step in identifying a suspected causal agent of disease or spoilage.

Curing A method of food preservation, with ancient origins, in which meat or fish is immersed in a solution of salt and potassium or sodium nitrate.

- Increased osmotic pressure creates conditions unfavourable to microbial growth by reducing the available water.
- During the process, nitrate is oxidised to nitrite which reacts with haemoglobin forming a pink-coloured substance (the characteristic colour of bacon and similar products).
- The traditional (Wiltshire cure) process takes some weeks and modern developments include 'in-slice' curing and tumbling, both of which achieve curing in a day.

- Some products may be smoked or have additions of sugar.

See: Corned beef, Halophile, Nitrate, Nitrite.

Custard A milk product and thus a high-risk food. It must either be kept hot (above 63 °C) or cold (below 8 °C). A danger is that custard may be prepared in advance, become contaminated and is then not reheated sufficiently.

Cutlery *See*: Cleaning schedules, Washing up.

Cutting board *See*: Chopping boards.

***Cyanobacteria* (Blue-green algae)** A group of micro-organisms which are subject to rapid multiplication in fresh water and are a known health risk to humans, fish, domestic and wild animals. Symptoms may prove fatal and include dizziness, headaches, muscle cramp, gastroenteritis and vomiting. Persons most at risk are those in direct contact with infected water. *See*: Algae.

Cyanogens Cyanide-containing substances found mainly in the seeds of foods including apples, pears, apricots and plums and also in various nuts and beans such as almond and kidney beans. Fatal poisoning has been noted following consumption of roasted apple pips. The tropical root vegetable manioc and the lima bean require thorough cooking to drive off cyanide as hydrogen cyanide gas.

Cysticercus bovis, cellulosae* and *tenuicollis The cystic stages of the beef tapeworm (*Taenia saginata*), the pork tapeworm (*Taenia solium*) and the dog tapeworm (*Taenia hydaginata*) respectively.

Dairy products The protein content of many dairy products makes them attractive to bacteria and they are often classed as being high-risk foods. Milk is subject to a great deal of food law such as the requirements for heat treatments of pasteurisation and sterilisation. It has been involved in outbreaks of both food poisoning and food-borne disease. Soft cheeses ripened with moulds or other micro-organisms are prone to growth of *Listeria* especially if raw milk has been used in their preparation. Cream has also been implicated in food poisoning outbreaks when cross-contamination has occurred. *See*: Butter, Cheese, Cream, Milk, Yoghurt.

Damaged stock Damage to packaging, containers and cans may allow food to become contaminated with micro-organisms or physical contaminants. Such foods should be regarded with suspicion and discarded if there is any doubt about their fitness. Food escaping after container damage may attract pests and also spoil other foods.

Danger zone This is the range of temperatures, 5 to 63 °C, in which food poisoning bacteria may multiply. Within the danger zone are:

- room (ambient) temperatures (often around 20 °C and above)
- human body temperature, 37 °C. This latter is the optimum or best temperature for the growth of many pathogens.

Under English law, many foods known as high-risk foods, must not be kept within the danger zone of temperatures. This is the principle of temperature control and includes the use of refrigeration, cooking and hot storage. If a hot, cooked item requires refrigeration it must not remain in the danger zone for more than 90 minutes. *See*: Appendix 1: Food Law, Cooling of food, Mesophile.

Dark mealworm (*Tenebrio obscurus*) A dull/matt black or brown insect, about 10 mm in length, which is a pest of stored cereal products including flour. It thrives in damp and dark conditions producing large (over 20 mm long) yellowish-brown larvae.

Dark-cutting meat A self-explanatory term used to describe meat of poor keeping quality, the colour being due to the muscle protein myoglobin. The most common cause is the slaughter of fatigued or stressed animals whose muscle glycogen reserves are low. This results in a high ultimate pH (over 6), glycogen being converted into lactic acid after death. 'Glazy' bacon has a similar cause.

Date codes Many foods are marked with dates indicating either when they were produced or their last date for consumption.

'Use-by' date
Foods which are perishable and could cause food poisoning have a 'use-by' date indicating the last day upon which the food is safe to eat. Examples include meat and dairy products and other high-risk foods.

'Best before' date
Foods which deteriorate in quality but which are unlikely to cause food poisoning may have a 'best-before' date indicating that taste, appearance and other quality or organoleptic factors are likely to change after this date. Examples include potato crisps, biscuits and canned drinks.

Other foods such as canned and dehydrated foods, bread, fresh fruit and vegetables do not have date codes. It is desirable to mark delivery dates on them where possible so that proper stock rotation can be carried out. Out-of-date stock should not be used or sold. If it causes problems then prosecution is almost inevitable. Manufacturers' codes are often present, with some systems being difficult to decipher.

DDT (Dichlorodiphenyl Trichlorethane) An insecticide developed in the late 1930s and widely used because of its toxicity to a large number of species. It was originally thought to be non-toxic to mammals but this has not proved to be the case. Higher life forms concentrate DDT in their tissues eventually building up lethal or damaging doses. Its use has been severely curtailed because of these essentially environmental effects.

Deadly nightshade A plant which occasionally causes acute poisoning.

Death Cap mushroom (*Amanita phalloides*) A highly poisonous mushroom. The appearance of it is unspectacular, being small and pale-coloured, and it is suspected that mistaken identity is the cause of at least some cases of poisoning. Symptoms appear up to 12 hours after ingestion. The grey squirrel is able to eat this variety without harm.

Death phase Alternative term for decline phase. *See*: Bacterial growth curve.

Decomposition/decomposers An indication of food spoilage in which the structure of cells is broken down by micro-organisms (decomposers) and enzymes to produce undesirable changes such as liquefaction and off-odours.

Deep freeze

Advantages

- At commercial deep freeze temperature of −18 °C and below, the activities of many micro-organisms are very much reduced, with multiplication being virtually non-existent.
- Shelf-life is greatly increased.

Disadvantages

- Deep freezing does not kill the majority of pathogens at temperatures commonly used.
- On thawing, bacteria will start to multiply as soon as the danger zone is reached.
- Spoilage enzymes are also inhibited at deep freeze temperatures but will continue to work slowly and eventually cause loss of quality.

See: Freezers (various entries), Spoilage of food, Star symbols.

Defaecate The act of expelling faeces from the human or other body. *See*: Intestinal flora.

Defrosting

1) In refrigeration equipment, ice tends to build up as a coating on

internal surfaces. This reduces the efficiency of refrigerators, freezers and chillers and should be periodically removed. Removal may be manual with a scraper or automatic using a built-in system. Automatic systems are designed so that the product does not reach an unacceptably high temperature.

2) An alternative term for thawing.

Dehydration

1) Loss of water from the human body as a result of continued vomiting and diarrhoea during an attack of food poisoning. It may be so severe as to cause death, and this has been noted in cases of salmonellosis.
2) A method of food preservation which uses various techniques to remove water from foodstuffs and so make them unsuitable for the growth of micro-organisms.

- Dried foods have a very long history, early products (c.3000 BC in Mediterranean regions) being produced by sun-drying; plums and grapes produced prunes and raisins; apricots and other fruits were also sun-dried.
- More modern methods include the spray-drying of milk and other liquids.
- Dehydrated foods may be kept in closed containers at room temperature for long periods but as soon as water is added (reconstitution or rehydration) they must be treated as if they were fresh foods. This is particularly important for high-risk foods such as milk, gravy and soups.

See: Accelerated Freeze Drying, Available water, High-risk foods.

Delivery of foodstuffs A vital part of the food chain and one in which cross-contamination and loss of proper temperature control is possible.

- Goods entering food areas must be carefully checked to ensure that they are received in a satisfactory condition before storage or use.
- Attention should be paid to date codes.
- Packaging and containers must be intact.
- Temperatures of chilled frozen foods should be checked.
- Avoid any contamination by keeping foods separate and not exposing them during delivery.
- All goods should be carefully checked against invoices and weighed if necessary.

- Vehicles and delivery persons must be hygienic.
- It is becoming increasingly common for detailed specifications of foods, including delivery conditions such as temperatures, to be demanded from suppliers.

Delivery vehicles

- Generally these and containers used in them must come up to the same standards of construction and hygiene as food rooms.
- The internal surfaces of delivery vehicles should be capable of being easily cleaned and food should not be subjected to any possibility of contamination.
- Segregation of foods should be possible if deliveries are of mixed foods.
- Certain temperatures may be required for the transport of chilled or frozen food; deliveries of such food may be rejected if temperatures are incorrect, for example chilled foods above 5 or 8 °C or frozen foods above −18 °C.
- Refrigeration and insulation of vehicles is normally required for high-risk foods.

See: Appendix 1: Food Law, Market stalls etc. Regs.

Department of Health (DoH) Government department which has major interests in food hygiene and food law. There are links between the Department of Health and MAFF (Ministry of Agriculture, Fisheries and Food), particularly if food safety is an emergency of national concern.

Dermatitis An often painful inflammation of the dermis of the skin which may be caused by contact with a sensitising substance. The skin cracks and infection may result. Many chemicals, including those used in cleaning and washing up have caused it, particularly through repeated use although cases have arisen from a single contact. Affected areas of skin must be protected to avoid both food contamination and making the condition worse. In severe cases, sufferers may not work in food areas.

Dermestid Common name for the *Dermestes* beetles. *See*: Bacon beetle, Hide beetle.

Desiccated coconut Dried coconut flesh is not normally associated with food poisoning. It is of historical interest because of outbreaks of typhoid/paratyphoid fever and salmonella food poisoning during the 1950s in many parts of the world, caused by contaminated coconut and its use in cakes and confectionery. Improved hygiene practices in the countries of origin of the product have proved effective.

Design of equipment in food premises
Must be such that it is:

- capable of being easily cleaned,
- non-toxic and impervious,
- corrosion-resistant,
- smooth, non-cracking and hard-wearing.

Stainless steel and certain plastic materials are often quoted as being ideal. Items of equipment should not cause physical contamination of food, for example from loose bolts. *See*: Appendix 1: Food Law, Chopping boards, Colour-coding, Copper, Metals, Polypropylene, Stainless steel, Wood, Zinc.

Design of food premises
- Mains services (such as gas and water) must be available.
- Regard must be given to location, bearing in mind planning restrictions and registration requirements.
- Buildings should be proofed against pests and comply with current legislation.
- The principle of linear workflow helps to minimise risk of cross-contamination.
- Similar requirements to materials used in food equipment apply to food premises.
- All surfaces such as floors, walls and work surfaces must be impervious and easy to clean efficiently.
- Facilities for staff changing and personal hygiene must be incorporated.
- Adequate storage facilities including refrigeration are essential.
- Adequate lighting and ventilation are required.

Desserts Those which contain milk and are above pH 4.5 must be kept refrigerated. Some yoghurts may be included. *See*: Appendix 1: Food law.

Destroying Angel mushroom (*Amanita verna/virosa*) A highly toxic mushroom closely related to the Death Cap mushroom.

Desulphotomaculum nigrificans A Gram-positive spoilage bacterium which causes sulphide or 'sulphur-stinker' spoilage in low acid heat-processed canned foods, such as peas and corn. It produces hydrogen sulphide (rotten eggs smell), without any discolouration in peas, but in corn there is some blackening of the product. The organism is fairly easily killed by heat, and its presence in the food is evidence of under-processing. There is no external evidence of spoilage as there is no swelling of the can. *Note*: The organism was formerly known as *Clostridium nigrificans*.

Detergents

- Cleaning agents used to remove grease and other substances from surfaces, utensils and equipment.
- They reduce surface tension and are described as surface active agents (surfactants).
- Grease is dispersed in water, suspended and prevented from resettling; this is known as emulsification.
- Detergents should not form scum; they should be non-toxic and be easily rinsed off.
- Common soapless detergents are anionic and have negatively-charged, water-liking (hydrophilic) heads.
- Cationic detergents (positively-charged head) are often bactericidal and used in sanitisers.

See: Amphoteric, Quaternary ammonium compounds, Soap, Soapless detergent.

Dettol One of a number of trade names for the antiseptic/disinfectant substance chlorxylenol. It has a strong taste and odour and has no use for food-contact items but may be used in other areas such as toilets and drains.

Diarrhetic shellfish poisoning (DSP) Caused by toxins released by the algae *Dinophysis* and *Prorocentrum* particularly in Japan, although the problem has spread to Europe. Gastroenteritis is the main symptom and since the toxins are heat-stable the best preventative measure is to monitor contamination of shellfish beds.

Diarrhoea A common symptom of food poisoning in which loose or liquid faeces are expelled with some frequency and violence. The cause is due to the presence of pathogens, toxins or other substances in the intestines. *See*: Food poisoning.

Diazinon An organo-phosphorus residual insecticide used in lacquers to control cockroaches, and in fly strips. It is toxic to mammals.

Dichlorvos A chlorinated organic insecticide used as a fumigant and in fly strips.

Diguanides Cationic detergents used in glass washing.

Dinoflagellates Unicellular, free-floating algae. Certain marine forms produce neurotoxins that cause paralytic shellfish poisoning. The disease spreads to humans when large numbers of dinoflagellates are eaten by molluscs (e.g. mussels), which in turn are eaten by humans.

Diphtheria An infectious disease caused by three types of the Gram-positive *Corynebacterium diphtheriae*, affecting the throat of humans. The bacterium produces a powerful toxin which affects the heart and nervous system and may be fatal. The disease is contracted by the respiratory routes or through wounds and skin infections and has been spread via milk contaminated by dairy workers.

Diplococcus Cocci that divide once and remain attached in pairs ('diplo' = two). *See*: Bacterial Classification.

Diptera Literally 'two winged', the rear pair having degenerated to rod-like structures. This order contains a large number of species of flies including the housefly and bluebottle, both of which are significant carriers of bacteria capable of causing food poisoning.

Diseases (in food handlers) As a general rule, all food handlers must report diseases or infections, which could be transmitted to other people, to their supervisor who should inform the local EHO. *See*: Exclusion of food handlers.

Dish-washing An important part of cleaning, the effectiveness of which is paramount to ensure that no food on which micro-organisms may grow is left on plates. The image of a food business is frequently judged by the cleanliness of crockery and utensils.

Two-sink system

1) Pre-clean or scrape to remove large food items.
2) Wash or main clean at around 55 °C with a detergent to remove grease; this low temperature prevents foods such as eggs being cooked onto the plates.
3) Disinfecting rinse in a second sink containing water at about 85 °C, or cooler water with a suitable chemical sanitiser; this removes detergent, reduces the bacterial load and heats up plates so that they may air dry without the use of cloths.
4) Air drying.
5) Finished items should be stored where they will not become contaminated.

- Lightly soiled articles such as glasses are washed first so that the detergent does not become contaminated straight away.
- Wash and rinse chemicals must be freshly made according to manufacturers' instructions and changed when necessary.

The single sink system is similar but requires filling and emptying. Mechanical dish washer cycles also resemble the two-sink system. *See*: Cleaning.

Disinfectant/Disinfection

- Disinfection is the reduction of the numbers of micro-organisms to a safe or acceptable level.
- It does not imply total destruction nor a significant reduction in the numbers of spores.
- Common disinfectants include heat (water at 82 °C applied for one minute is a common use), various chemicals such as hypochlorites and hydrogen peroxide and ultra violet light.
- Articles which are disinfected in the food industry include machinery, food contact surfaces, cleaning equipment and the hands of food handlers.
- Chemical disinfectants must be made up and used according to manufacturers' instructions and not kept for long periods since they become inactive.
- They require time to work; up to 30 minutes contact may be needed.
- Soap and hot water is often stated to be the best and cheapest disinfectant and is particularly useful in hand washing.

Dispensers (soap) Tablets or bars of soap are known to support bacterial growth and liquid soap in a dispenser is preferred. The soap may be germicidal and should be non-perfumed to avoid taint.

Dogs

- Neither these, nor any other live animal, should be allowed in food rooms due to the wide range of diseases they may carry.
- There is also danger of physical contamination of food from hairs or any substance present on the feet or fur of the animal.
- Local byelaws exist which prevent members of the public from taking dogs into food premises (guide dogs excepted).
- Examples of canine-transmitted diseases include leptospirosis canicola, hydatid tapeworm, fungal infections, salmonellosis and toxocariasis.

DoH *See*: Department of Health

Domestic premises Increasing use is made of domestic kitchens to prepare food for functions and cases of food poisoning have occurred which involved food thus prepared.

Such premises may be unsatisfactory because:

- equipment is likely not to be made to handle the heavy use for which commercial units are designed;
- refrigeration may be inadequate;
- the layout of the kitchen will not be so as to minimise cross-contamination;

- separate hand-washing and equipment sinks are not normally present;
- the room is used by non-food handlers and pets.

See: Design of equipment/ food premises.

Doner kebab A spit-roasted minced meat product (lamb or sometimes beef), sometimes hand-moulded before cooking, and removed in slices by carving.

- Correct temperature control during preparation and cooking is vital since bacteria present will be all through the kebab and not solely on the outside.
- Destruction of pathogens depends on thorough cooking and a high temperature must be achieved at the centre, making sure the the grill is not turned down or off at any time.
- Any left-overs must not be reheated.
- Since the product is handled during preparation, personal hygiene must be of a high standard and raw kebabs must be refrigerated before use.

Care should also be taken with salad materials used, especially since they are often prepared in bulk in advance.

Door handles Often contaminated after contact with hands, a particular example being the internal door handle in a WC compartment. Handles should be included in cleaning schedules.

Doors Like all other surfaces in food rooms, doors should be finished with an impervious (non absorbent) surface so that they may be readily cleaned. Kick plates may be necessary at the bottom to prevent access to rodents and the door should fit well leaving no gaps.

Doubling time Alternative term for generation time.

Drainage An essential service requirement for all premises. It needs careful design and planning (in consultation with the local authority) to ensure conformation with building regulations.

- Grease traps may need to be installed to prevent blockage of the system by congealed fats; they need periodic emptying and are best located outside food areas.
- All sinks and sanitary equipment are connected to the drainage system via a trap, the function of which is to prevent odours entering the room and to provide a barrier against rodents.
- Some items of equipment such as peeling machinery may have filter traps to prevent large quantities of debris entering the drainage system.

- Floors in food rooms are often laid with a gentle slope (1 in 25) to a half-round drainage channel, which is covered by a grating, or alternatively to a trapped inlet.

Dried foods *See*: Dehydration.

Drip Water lost from frozen food as it thaws, resulting in loss of quality since the liquid lost originated inside cells.

- The amount of drip is proportional to the time taken to freeze the food, quick freezing in under one hour producing least drip.
- In meats quick freezing time is not feasible for large items (such as entire carcases) although a high ultimate pH reduces drip considerably.
- Smaller food items may be frozen very rapidly and tend to suffer less drip.

Drinking water Must be provided for employees at all places of work including food businesses. The responsibility for ensuring that water is fit to drink (potable) rests with the water supply company. After it has left the tap in a food business it becomes a food under the definition of the Food Safety Act 1990.

Drinks Subject to much the same hygiene problems as foodstuffs particularly where there are high-risk components such as milk. Water is a known vehicle of food-borne diseases such as typhoid fever.

Droplet infection The transmission of infection by small liquid droplets which carry micro-organisms. The liquid (mucus, from the nose, mouth or throat), will carry bacteria, and if allowed to contaminate open food, may result in an outbreak of food poisoning. *See*: Personal hygiene, Sneeze screens, *Staphylococcus aureus*.

Droppings The faeces of rats, mice, insects and other animals. These are often the first visible sign of infestations and have characteristic sizes and shapes from which identification of pests may be made, e.g. the droppings of the black rat have a curved appearance. The presence of different sizes of rat droppings usually means that the infestation involves several generations.

***Drosophila* (fruit fly)** Small flies which feed on decaying fruit, sour milk and fermenting foods such as wine and vinegar, apparently being attracted by smell. There are several varieties including *D. melanogaster*, *D. funebris* and *D. repleta*, all of which lay large numbers of eggs which become maggots and eventually pupae before the adult flies emerge.

Drugstore beetle (*Stegobium paniceum*) A winter-hardy pest of stored cereal products and pharmacies, also called the biscuit beetle. It is a small, reddish-brown insect with a hooded thorax and is covered with short yellow/brown hair. It frequently infests domestic premises with the larvae (brown, 5 mm long) burrowing through packaging materials to infest food.

Dry goods stores Non-perishable foods and some preserved foods such as cans and dehydrated products are best kept in a well-ventilated room at between 10–15 °C.

- The construction and equipment used should be of the same standard as for food rooms.
- In order to allow effective cleaning to take place, storage should be off the floor.
- A stock rotation system must be used and sound containers used to prevent pest infestation.
- Humidity is best at around RH 60%.
- Store rooms should be well-lit and ventilated.

Drying (after washing up) If the final rinse temperature is hot enough (over 82 °C) then air drying may be carried out. This avoids the use of potentially contaminated cloths. *See*: Cleaning, Washing up.

Drying of hands Suitable facilities for this are required by law so that the important personal hygiene function of hand-washing may be completed. Several methods exist, including hot air, single use (disposable) and roller towels, all of which have advantages and disadvantages.

1) Hot air dryers, if used correctly are very hygienic but tend to be slower than other methods, causing misuse and non-use if a queue develops. A back-up system using disposable towels is often installed in case the dryer fails.
2) Disposable paper towels are very effective as long as the supply is sufficient and disposal is satisfactory.
3) Mechanical roller towels are good as long as they are replaced as required and the mechanism does not malfunction.
4) A length of ordinary towelling on a roller quickly becomes contaminated and is not suitable for use by many people.

Dublin Bay prawn (*Nephrops norvegicus*) Also called the Norway lobster, langoustine or scampi. A large (20 cm) lobster-like crustacean coloured orange/pink after boiling. *See*: Shellfish.

Duck eggs Known to have been the vehicle of species of *Salmonella* in a number of cases of food poisoning. The shell is porous and micro-organisms may penetrate before and after laying. Raw or lightly cooked duck eggs are not generally regarded as being safe. *See*: Eggs, shell (duck).

Ducks *See*: Poultry.

Due diligence A legal term under the Food Safety Act 1990. It indicates the need for producers of food to take all reasonable precautions to avoid causing food safety offences.

- Doing nothing is not enough – some positive action is required.
- A system of controls must be set up, implemented and monitored; this includes:
  Hygiene of staff, premises and equipment
  Specifications for raw materials
  All preparation and production operations
  Compliance with all legal standards
  Labelling, description and advertising
  Staff training
  Dealing with customer complaints.

Manufacturers frequently use BS 5750, ISO 9000 or other quality assurance.

- All precautions which can reasonably be taken must be taken.
- Circumstances dictate what is reasonable. For example a large cook–chill production unit has to take more precautions than a small sweet shop because of the far greater risks involved.
- Written records should be kept.

If it can be shown that another person (not under the control of the person running the food business) has caused the problem, by either an action or by default, due diligence may be proved. The same would be the case if an alleged offence has been committed despite all reasonable precautions having being taken. There is a guidance booklet on due diligence prepared by IEHO, FDF, LACOTS and others.

Duodenum The first part of the small intestine.

Dust May be both a physical contaminant of foods and also a source of soil-borne organisms such as *Clostridium* and *Bacillus* species, both of which are frequent causes of outbreaks of food poisoning.

Dustbins *See*: Refuse (storage and disposal), Waste (control and disposal of).

Dysentery (amoebic/bacterial) Inflammation of the large intestine resulting in diarrhoea and other symptoms including nausea and pain. There are two types of organism responsible for the disease: bacterial and protozoan. *See*: Amoebic dysentery (amoebiasis), Bacillary (bacterial) dysentery, *Entamoeba histolytica, Shigella.*

EC (European Community) An economic/political organisation of twelve European countries including France, Germany and the UK. It has developed from a trading group and generally has increasing powers over a wide range of activities in its member states. The aim of at least some members is to eventually achieve both economic and political union in a 'United States of Europe'. In terms of food hygiene and related matters this could mean that standards would eventually be similar across the whole community, for example in composition of foods, labelling, the use of additives and general hygiene during food production. Many EC directives and regulations have been issued with this end in view but harmonisation is likely to take some years.

EC Directive The European Community issues directives on matters such as food hygiene, safety at work and trade, with which member states comply by legislation. The overall aim is to ensure similar regulation in all member countries by certain dates.

***Echinococcus granulosus/veterinorum* (Hydatid disease)** A tapeworm parasite of dogs whose cystic stage may infect humans and food animals. Many human cases are caused by eating food contaminated with dog faeces.

E. coli A commonly used abbreviation for the bacterium *Escherichia coli.*

Eggs, liquid (whole) Bulk liquid egg is rendered safe by pasteurisation. However it needs to have refrigerated storage and protection from contamination, since it is a product which readily supports the growth of micro-organisms.

Eggs, shell (duck) Duck eggs are known to contain significant numbers of salmonella organisms.

- Ducks live in and around ponds with their nests nearby, and so their nests tend to be in muddy places; the high moisture environment favours penetration of the shell by micro-organisms.
- Shells of duck eggs are porous when laid unlike those of hens' eggs which have a waxy coating.

- The duck is a scavenger and it shares its surroundings with many other wild animals, possibly including rats.
- There is a high chance that the bird itself may become infected with a variety of bacteria, including species of *Salmonella*.

Use of duck eggs

- They must be treated with caution.
- Refrigerated storage is essential.
- They should not be used in dishes which require no or only light cooking, such as in mousse or omelettes.
- Lightly boiled or lightly fried eggs must be avoided.
- Young children, the elderly, people who are ill, and pregnant women are particularly at risk.

Eggs, shell (hen)

- Until the late 1980s, raw hens' eggs were considered to be a safe food, provided the shell remained intact. There was always the chance that contamination from the outside of the shell surface (e.g. from faecal contamination from the hen), or elsewhere (e.g. through washing or packaging) could occur, if the shell was cracked or during the act of breaking.
- In the late 1980s a number of outbreaks of food poisoning occurred, which were associated with the consumption of lightly cooked eggs.
- The bacterium *Salmonella enteritidis phage type 4* was found to be the organism responsible; it was found in the yolk of unbroken eggs and it is now known that it is possible for the organism to be in the egg when laid.
- Such contaminated eggs are produced by birds which themselves contain the organism.
- The feeding meal given to the birds is thought to have been responsible for the infections; heat pasteurisation of such meal is considered sufficient to destroy the organisms.

Use of eggs

1) Eggs with cracked shells should be avoided.
2) Refrigerate during storage.
3) Do not use raw egg for the production of foods which are not to be subsequently cooked, such as mayonnaise or mousse. Commercially produced mayonnaise made from pasteurised eggs is available.
4) Young children, pregnant women and the elderly or ill should refrain from eating lightly cooked eggs.

Eggs (spoilage) Some of the following spoilage effects are noticeable with the naked eye, others are only shown by candling.

1) Fresh eggs
   Cracks in the shell, leaks, stained or dirty spots on the outside, blood clots, and clear spots in the yolk (the latter two can only be seen through candling).

2) During storage
   a) Microbial changes to an undamaged egg require initial penetration of both the shell and internal membranes. Bacterial spoilage may take up to several weeks at refrigerated temperatures.
   - Effects include various 'rots', caused by a large variety of bacteria (e.g. green, colourless and black rot); they all produce 'off' flavours, and the black rot produces the putrid, foul-smelling 'rotten egg' smell (caused by *Pseudomonas fluorescens*, *Alcaligenes* and *Proteus* species).
   - Fungal spoilage includes 'pin spot' moulding of various colours (yellow, green, blue, black and pink) on the surface of the shell and just inside it. Such spoilage moulds include *Penicillium*, *Cladosporium*, *Mucor* and *Sporotrichum* species.
   - This is often followed by further fuzzy mycelial growth on the surface, followed finally by fungal rotting after the mycelium has penetrated the egg through the pores or cracks in the shell, typically by *Cladosporium* and *Sporotrichum*.

   Eggs will be more likely to spoil if moisture is allowed to accumulate on the shell. This may be prevented by maintaining a constant temperature during storage.

   b) Non-microbial spoilage includes the loss of moisture, leading to weight loss, thinning of the white of the egg, and weakening of the egg yolk membrane. When opened, affected eggs show thinned whites and the yolk may flatten or even break.

Egg white (albumen) This is not a good medium for microbial growth due to the following reasons:

- the low content of non-protein nitrogen;
- the unavailability of certain vitamins through combination with proteins;

- the lack of 'free' iron, needed by micro-organisms for enzyme activity;
- its high pH of 9.6.

All of these protect eggs from microbial attack during proper storage.

Egg yolk Egg yolk used for glazing pastry readily supports the growth of micro-organisms. It requires refrigeration when not in use, and care needs to be taken to protect it from cross-contamination, and to prevent it from contaminating other foods, surfaces and equipment. Ideally it ought not be retained for more than two or three days at a time.

Electromagnetic radiation Radiation including X-rays, gamma rays (food irradiation), ultra violet (flying insect killers), infra red (some cooking equipment), visible light and radio waves.

Emergency prohibition notice and order Under the Food Safety Act 1990:

- If the condition of a food premises is thought to be an imminent health risk it may be closed immediately on the judgement of an EHO or other authorised officer.
- A notice to this effect is conspicuously displayed on the premises.
- The closure must be confirmed by court order following application to the court within three days or compensation may be payable.

Common reasons for issue of such orders include:

- pest infestations
- very dirty conditions.

Other legal action may also follow. The order remains in place until the enforcement authority issue a certificate stating that the health risk has been eliminated. The person running the business must apply for this certificate. There is a right of appeal at all stages.

Emulsion A suspension of small droplets of a liquid in another liquid. Butter is a water-in-oil emulsion while water after washing up is an oil-in-water emulsion.

Enamel Metal or ceramic ware coated with a glaze which contains toxic chemicals such as antimony or cadmium. It may cause chemical food poisoning if used with acid foods.

Endospore

- A non-growing, heat-resistant structure that some bacteria are able to form inside the vegetative (growing or multiplying) cell.

- They are formed when environmental conditions become unfavourable, such as very high temperatures, lack of nutrients or low moisture.
- They can withstand strong chemicals, such as acids, disinfectants, and the destructive effects of radiation.
- The endospores of two groups of bacteria, the *Bacillus* and *Clostridium* groups, are the most heat-resistant of all.
- Endospores may survive for many years, for example in the soil, as the endospore coats make them resistant to dehydration.
- The most important potential danger associated with endospores in food preparation concerns their ability to withstand most normal cooking procedures.
- They can for example survive for up to five hours in boiling water. In comparison, vegetative cells are destroyed in a few minutes at this temperature.
- Any spore-formers, such as *Bacillus cereus* and *Clostridium perfringens* will form endospores as soon as the temperature starts to rise significantly.
- They will survive cooking unharmed and will revert to the vegetative stage in a matter of minutes as soon as the temperature falls to below 63 °C, so careful treatment of foods after cooking is essential.

Endotoxins

- Poisonous substances released from bacterial cells and some other micro-organisms such as fungi.
- They are not released from the cell during its life, but are released from cells only after the cell wall has been damaged or following death and subsequent fragmentation of the cell (e.g. in the intestinal tract), or during spore formation.
- Endotoxins may take some time to cause symptoms in the human body since it takes some time for their release.
- Typically they are found in Gram-negative bacteria, such as *Escherichia coli* and *Salmonella* but also in *Clostridium perfringens* (a Gram-positive bacterium).
- Endotoxins are not easily destroyed by heat and are responsible for such symptoms as diarrhoea, local tissue shock and fever.
- Aflatoxin (a mycotoxin) is an endotoxin produced by some fungi.
- Endotoxins are conjugated proteins (i.e. proteins combined with a non-protein molecule), and are part of the structure of the cell, usually part of the outer membrane.

End-seam In canning, the join between the lids (end plates) and body. The dimensions of the seam are important. *See*: Canning.

Energy in cleaning Three types of energy are involved in most cleaning methods.

1) Kinetic or physical energy provided by machine or person.
2) Heat energy, normally supplied by hot water.
3) Chemical energy, supplied by detergents, sanitisers and disinfectants.

See: Cleaning.

Entamoeba histolytica A single-celled organism (a protozoan) which causes the water-borne and food-borne disease known as amoebic dysentery. It frequently causes ulceration and bleeding of the intestinal wall, with blood and slimy mucus in the faeces, fever and chill. Within the intestine, and in the faeces of affected individuals the disease exists in two forms.

1) The amoeboid (vegetative) state, which is easily destroyed when released from the body (and so is unlikely to spread the disease).
2) A resistant cyst form, that is easily transmitted in water, food, or even by houseflies. It is not destroyed by the chlorine used in the water treatment process. A new infection follows the swallowing of cysts in contaminated food or water, which set up an intestinal infection by then taking up their amoeboid form.

- The severity of the disease depends on the number of intestinal lesions produced and the extent of damage to the lining of the intestine.
- Many victims lack any symptoms and so act as carriers of the disease.
- Those with symptoms may have mild, but prolonged diarrhoea or they may have moderate diarrhoea containing blood and mucus. Rarely, abscesses may occur in the liver and/or other organs.
- The disease may last for months; it sometimes flares up after travellers have returned from countries where the disease is common, such as SE Asia, parts of Africa, China and Mexico.
- There is a significant reservoir of infection in the United States.
- Drug treatment is required and is effective.
- Good public health measures and efficient sewage treatment control the disease.

Enteric fever A general term used to describe both typhoid fever and paratyphoid fever, both of which are exclusively diseases of man.

Enterics *See*: Enterobacteriaceae.

Enteric viruses *See*: Enterovirus.

Enteritis Inflammation of the intestines, sometimes a symptom of food poisoning.

Enterobacter aerogenes A Gram-negative bacterium, which is one of the coliform group of organisms; it is found in the intestinal tract and is an opportunistic pathogen. It is also commonly of plant origin. This organism, together with *Escherichia coli* and *Klebsiella*, are frequently used as indicators of faecal-contaminated water. *Note*: This organism was formerly known as *Bacillus aerogenes* and later as *Aerobacter aerogenes*.

Enterobacteriaceae A group of Gram-negative bacteria that inhabit the gastro-intestinal tract of man and animals. Some are resident organisms and others are present only when causing disease. Included in this group are *Escherichia coli*, and the *Enterobacter*, *Shigella* and *Salmonella* groups. *See*: Coliforms, Individual named organisms.

Enterotoxins

- A general term for poisonous waste products which act on the lining of the intestine or stomach, causing diarrhoea or vomiting.
- They are fairly resistant to heat, surviving for 30 minutes at 100 °C. The bacteria which produce them are often much less heat-resistant (e.g. *Staphylococcus aureus*) and their presence in contaminated food is not essential for illness to occur.
- Endotoxins and exotoxins are both examples of enterotoxins.

Enterovirus A term used to describe any virus which can be transmitted by faeces. They can be spread from person to person by direct contact or by water and food contaminated by faeces. They may be found in both fresh water and sea water. Included in this group are the viruses that cause poliomyelitis and hepatitis (HAV).

Entry (power of) *See*: Power of entry.

Environmental Health Officer (EHO) An authorised officer of a local authority who is empowered to enforce various laws and perform certain functions, including checks on food hygiene, health and safety provisions, pollution control, housing, infectious disease control and many others.

- The modern EHO was previously called Inspector of Nuisances, Sanitary Inspector and Public Health Inspector and is required by law to have a professional qualification (diploma or degree) before appointment. They become corporate members of the Institution of

Environmental Health Officers following an assessment of their professional competence. Most local authorities will only appoint such persons as 'authorised officers' under the Food Safety Act, 1990.
- The EHO on occasions works with HSE inspectors and Trading Standards Officers.
- They have power of entry and may visit premises at any reasonable time, take photographs, samples, etc.

Enzymes Protein-based, biological catalysts, which have great importance in the biochemistry of all living organisms. After harvest or slaughter, enzyme action continues.

- They can cause food spoilage and decay.
- Most work best at body temperatures and neutral pH.
- Their action decreases as temperature falls or rises.
- They continue to work in refrigerators (1–4 °C).
- They work very slowly in freezers (−18 °C).
- They are destroyed (denatured) by heat, e.g. as a result of blanching or cooking.
- A low pH denatures them e.g. in vinegar.

Note: A catalyst speeds up the rate of a reaction without itself being affected, so it may be re-used again and again.

Epidemiology The study of environmental and other factors, with regard to when and where diseases occur and how they are transmitted in human populations.

Equipment (design and construction) Equipment likely to come into contact with food must be made so that it can be effectively cleaned. Food grade stainless steel is an ideal material for many items of machinery as long as the method of cleaning is satisfactory. *See*: Chopping boards.

Ergotism A fungal disease in humans caused by eating rye contaminated with an alkaloid mycotoxin released by *Claviceps purpurea*. Symptoms include convulsions, hallucinations and may result in death. The disease is also known as St Anthony's fire. *See*: Aflatoxin.

Erwinia Food spoilage bacteria causing rot in fruits and vegetables.

***Escherichia coli* (*E. coli*)** Found in the gut of man, farm animals and domestic pets. This bacterium does not normally cause disease, but certain strains may cause illness due to:

1) the release of an enterotoxin, which causes a cholera-like illness

2) invasion of the intestinal lining, causing a dysentery-like illness
3) a different mechanism that is not understood.

- It is this third group which are a common cause of 'traveller's diarrhoea', following consumption of *E. coli*-contaminated food or water.
- It causes food poisoning by infection, with an incubation period of 12 to 72 hours, with the illness lasting from one to seven days.
- Symptoms include abdominal pain, watery diarrhoea, vomiting and sometimes fever. It usually requires a large infective dose, therefore requiring poor storage and temperature control after contamination, or a high initial contamination dose.
- It grows best between 10 and 40 °C.
- It is commonly found in raw foods of animal origin.
- Foods commonly involved are meat, poultry and fish, and their products.
- It is not a spore-former, so it is easily destroyed by heat, e.g. by proper cooking of foods, or by pasteurisation.
- Generally it is more of a problem for infants and young children. Most of the severe outbreaks have occurred in hospitals, nursing homes and nurseries.
- For infants, breast feeding is the best preventative measure.
- Strains of *E. coli* are also able to cause food spoilage.
- A Gram-negative coliform organism.

Control involves cooking food thoroughly, chilling food rapidly in small quantities, good personal hygiene, and efficient sewage and water treatment.

Ethanoic acid *See*: Acetic (ethanoic) acid.

Eukaryotic A type of cell possessing a true nucleus, separated from the surrounding cytoplasm by a nuclear membrane. It is possessed by the single-celled protozoa, the algae, fungi, plants and animals.

Evaporated milk A canned product similar to condensed milk, but having no sugar added.

Exclusion of food handlers

- Food handlers suffering from food poisoning should not be allowed to enter food rooms until medical clearance is obtained.
- In the case of staphylococcal food poisoning this may be when symptoms have subsided.
- In the case of *Salmonella* infections, only after three consecutive daily faecal samples have proved negative.

- Any confirmed typhoid cases face a lifetime exclusion.
- Many other conditions such as heavy colds, boils, septic cuts, whitlows and styes may also warrant exclusion, each case being judged on its merits.

Excrement Term commonly used to mean faeces.

Excreta The normal discharges from the body, i.e. faeces, urine and sweat.

Excreter An individual who, whether suffering from a disease or not, persistently passes disease-causing organisms from the body in the faeces or urine. *See*: Carrier.

Exotoxins

- Poisonous substances released by micro-organisms such as *Staphylococcus aureus*, *Bacillus cereus* and *Clostridium botulinum*.
- They are actively released from the bacterial cell during its growth in food before its consumption, often producing symptoms within hours of eating contaminated food.
- Most exotoxins can be destroyed by heat; e.g. the exotoxin of *Clostridium botulinum* (which causes the disease known as botulism and affects the nervous system) is destroyed in 30 minutes at 60 °C.
- The exotoxin of *Staphylococcus aureus* (which affects the digestive system) is responsible for many outbreaks of bacterial food poisoning and can survive for over 30 minutes at 100 °C.

Tissue damage can be severe and at a great distance from the bacterial cells which produced it. Exotoxins are proteins produced by the cytoplasm of bacteria, both Gram-positive and Gram-negative cells.

Exponential growth The increase in numbers of cells by a constant doubling of cell numbers, i.e. 1, 2, 4, 8, 16, 32, 64, 128, 256 and so on. Such an increase in numbers leads to extremely large population numbers in a fairly short period of time. As a rough guide one bacterium will easily produce one billion bacteria in ten hours. It is also known as logarithmic growth.

Extraction systems (food room use) Hoods or canopies placed over cooking equipment collect fumes and water vapour which is extracted through ducting by fans. Filters may be provided to trap grease which would otherwise build up in the inaccessible interiors of ducting. Deep cleaning of ducts is essential periodically. Design of ventilation and air conditioning systems is an expert task. *See*: Ventilation.

Factory Inspector Enforcement officer appointed by the Health and Safety Executive (HSE) having duties under a large variety of safety law.

Facultative A term used to describe the ability of a micro-organism to grow under a variety of conditions. Facultative means optional and the term is often used to indicate the ability of a micro-organism to live in both the presence and absence of oxygen.

Faecal-oral route A common way of spreading disease through faecal matter carried on the hands. Such matter, which will certainly contain many forms of bacteria, and possibly protozoa, can easily contaminate food. In young children, it can be ingested directly, through sucking fingers, etc. As a means of spreading disease, it can easily be prevented by good personal hygiene, particularly hand-washing (with soap, and hot water) after using the lavatory, changing or handling babies' nappies etc. There are many diseases which are frequently spread by this means, including amoebic dysentery, bacillary dysentery (shigellosis), cholera, paratyphoid and typhoid fever and poliomyelitis.

Faeces Excrement discharged from the bowel, normally semi-solid in consistency. It is a ready source of bacteria (including pathogens), protozoa, and the eggs of parasites including tapeworms. Insects, rodents and shellfish feed on faeces, ingesting organisms which may be passed on, possibly contaminating food.

Faggot A meat product made from spiced chopped liver, spleen and lung (offal) with cereal binder. It is wrapped in a fatty abdominal membrane (omentum) and cooked. Since it is a high-risk food it must be subject to strict temperature control and treated in the same way as other meat products and cooked meats.

Family outbreak of food poisoning Two or more cases in the same household not connected with any other cases.

Fasciola hepatica A liver fluke affecting many animals including cattle, sheep and humans. Symptoms in humans are non-specific but the livers of food animals may be unfit if the infestation has badly thickened the bile ducts (so-called 'pipy' liver). Cases of the disease fascioliasis have been reported after eating wild watercress grown in water contaminated with sheep or cattle faeces.

Fascioliasis The disease caused by the liver fluke, *Fasciola hepatica*.

Fats

- Substances found in varying quantities in a number of foodstuffs, including meat, butter and margarine.

- Fats do not support the growth of bacteria but are prone to develop off-flavours and odours during storage, thus becoming unfit.
- They are chemically similar to oils having three fatty acids attached to a glycerol 'backbone' molecule. The major difference between fats and oils is their melting points.
- Saturated fats (few double bonds) tend to be hard animal fats, while unsaturated and polyunsaturated fats tend to be of vegetable or marine origin. *See*: Fatty acid, Rancidity.

Fatty acid A hydrocarbon chain, up to 35 carbons in length, ending in a carboxyl group, – COOH. The bonds between carbons may be single or double; the more double bonds the more unsaturated the chain. Unsaturated fatty acids are more reactive and foods containing these may spoil faster than those with saturated fats, hence, for example frozen beef has a longer life than frozen pork. *See*: Hydrogenation, Rancidity, Smoke point.

Favism An anaemic disorder caused by consumption of undercooked broad beans (fava beans), or an allergic reaction to its pollen.

Feeding stuffs Animal feeds have been known to be contaminated with various pathogens. The practice of recycling carcase meat as feedstock, if it has not been sterilised, has been proved to be responsible. Contaminated feed has been held partly responsible for the increase in *Salmonella* infection in intensively reared poultry. *See*: BSE.

Fenitrothion A general purpose organophosphorus substance often used as a residual insecticide to control cockroaches.

Feral Means 'wild' and is applied to pigeons and cats to distinguish them from domesticated individuals.

Fermentation

1) A chemical reaction involving the conversion of sugar into alcohol by the action of yeasts (*Saccharomyces*) and accompanied by the release of carbon dioxide gas. Mainly found in wine-making and brewing although fruits may ferment by the action of naturally-occuring yeasts.
2) A preservation process in which lactic acid is produced by various micro-organisms. Sauerkraut and yoghurt are typical products. The flavour enhancer, monosodium glutamate is also produced by bacterial fermentation of soya beans.
3) A laboratory method of identifying bacteria such as *Clostridia*.

Fever A bodily reaction to infection involving a rise in temperature.

Fimbriae Small unbranched extracellular appendages possessed by the bacteria. There are two forms, long and short. The long fimbriae are involved in sexual reproduction; the short are more numerous and are thought to enable bacteria to stick to a surface or to other cells. This latter property is related to the ability of bacteria to cause disease. Fimbriae are also known as pili.

Fines The Food Safety Act 1990 has increased the maximum penalties available to the courts. Magistrates Courts can impose fines of up to £2 000 per offence and a prison sentence of up to six months. For offences relating to the sale of unfit or injurious food Magistrates Courts can fine up to £20 000 per offence. Cases tried by crown courts can result in prison sentences of up to two years and the imposition of unlimited fines. *See*: Food Safety Act 1990.

Fingernails *See*: Nails.

Finnan haddock A cured and smoked fish, often dyed yellow.

Firebrats (*Thermobia domestica*) A greyish speckled wingless insect with long antennae, a relative of the silverfish (*Lepisma saccharina*). It infests bakeries and kitchens and other warm places, feeding mainly on cereals (also crumbs and other food scraps). This insect has the odd habit of biting holes in inedible fabrics. There is no evidence of it carrying pathogens, but it may become a nuisance (e.g. physical contaminant) rather than a serious health risk.

First Aid

- Food Hygiene (General) Regs 1970 require that suitable and sufficient first aid materials must be provided in all workplaces including food businesses, the exact amount being determined by the numbers of people employed and the nature of work done.
- An important item required by law in a first aid box is a supply of waterproof dressings.
- Trained first-aiders and sometimes medical staff are also needed where large numbers of people are employed.

Fish

- As protein-containing foods, fish and shellfish are included in the high-risk category and must be subjected to strict temperature control and proper handling.

- Raw fish and shellfish must be separated from other foods to avoid cross-contamination.
- Fresh fish deteriorates rapidly and may be stored at 0 °C, preferably in ice, for periods of up to two weeks depending on the variety.
- Cooked fish or shellfish must either be refrigerated (1 to 4 °C) or kept hot (above 63 °C).
- Unfit fish have a dull lustreless surface appearance often with beads of mucus, dark coloured blood around the gills, sunken eyes, limp bodies and an 'off' smell due to the production of trimethylamine (TMA).
- Certain species such as skate and dogfish produce a characteristic smell of ammonia early in decomposition.
- Several parasites (including tapeworms) are carried in fish, but these are normally destroyed when fish is cooked, smoked or frozen.
- Fish which is to be eaten raw should be frozen at −18 °C for at least 24 hours.

Fish and shellfish are capable of carrying many parasites, pathogenic bacteria and other poisonous substances including:

Anisalcis simplex, Chloromyscium thyrsites, Ciguatera, *Clostridium botulinum*, *Dibothriocephalus latum*, *Escherichia coli*, Mercury, *Phocanema*, Sarcotases, Scombrotoxin (affecting the Scombroid species – tuna, mackerel, etc.) and *Vibrio parahaemolyticus*. *See also*: Furunculosis, Paralytic shellfish poisoning (PSP). Details are under individual entries.

Flagella (singular flagellum)

- One of the two unbranched thread-like extracellular appendages found in the bacteria (the other being fimbriae).
- Many bacteria are capable of active movement (actively motile) in liquid surroundings, this being achieved by the possession of these structures.
- They are composed of a special contractile protein (flagellin) and are so small (20 to 30 micrometres in length), that special staining techniques must be employed in order to see them under the light microscope.
- The flagellin is arranged in a coiled helix and each flagellum is attached to the cell, inside the cell wall, to a basal body. These are thought to initiate and coordinate contractions of the flagella.
- In movement, the flagella rotate like propellors, and pull the bacterium through the water in a helical path.
- The number and distribution of flagella varies amongst the bacteria.

Some forms possess a single flagellum only, some have them distributed over the whole cell while others have one or more tufts of them.
- They are commonly found in the bacillus and spirillum forms.
- Some non-flagellate forms can also demonstrate motility, this being achieved by flexing of the cell.

Flat sours Organisms such as *Bacillus stearothermophilus* and *Bacillus coagulans*, which attack carbohydrates in low and medium acid canned foods (such as meats, soups and vegetables) cause spoilage ('sours') but not gas production ('flat'). Since no gas is produced the can vacuum is unaffected. These bacteria are both thermophiles and facultative anaerobes. Slow cooling of foods after heat treatment produces favourable temperatures for their growth (50 to 70 °C), as does storage of cans at high ambient temperatures. *See*: Canning.

Flavobacterium A group (genus) of Gram-negative bacteria commonly found in the soil. They may be responsible for contamination of water supplies and are spoilage agents of a wide range of foods, including beer (during production), shellfish, poultry, eggs, butter and milk, and they may cause discolouration on the surface of meats. They are obligate aerobes and some members are psychrotrophic and so are able to grow at refrigeration temperatures. Some have been found growing on thawing vegetables.

Fleas Small insects capable of jumping great distances; they are parasitic on warm-blooded animals. Over 1000 species are known, examples being the dog flea (*Ctenocephalides canis*), the cat flea (*Ct. felis*) and the human flea (*Pulex irritans*). Most fleas will attack other hosts, of particular note being the rat flea (*Xenopsylla cheopis*) which is the vector of plague.

Flesh fly (*Sarcophoga carnaria*) A grey-coloured blowfly with a chequered thorax sometimes infesting refuse bins.

Flies Insects belonging to the order of Diptera (two wings) and affecting foods in many ways.

- The families *Muscidae* (housefly) and *Calliphoridae* (blowfly) are most important since they are vectors of disease and cause food spoilage.
- Flies regurgitate (vomit) partly-digested food during feeding and may transmit bacteria to food in this way.
- They can only feed on liquid food and use the regurgitation technique to liquify solid food.
- Maggots developing from eggs grow in many foods, particularly meat.
- The characteristic 'fly spots' seen where flies have been feeding are the dried liquid faeces of the insects.

- All types may carry bacteria on their feet and bodies.
- Physical contamination of food is possible with bodies, pupae, egg cases and fragments of flies.
- The fruit fly (*Drosophila*) feeds on sweet foods and fermenting wines and beers.

Control of flies and other flying insects can be achieved by:

- screening open windows with mesh
- using plastic strip doors
- keeping refuse bins lidded
- keeping storage areas tidy
- using ultra-violet flying-insect killers
- using fly-papers away from food.

The use of insecticides in food rooms is a specialist task; fly sprays are not recommended because of the danger of food contamination.

Floors As with other parts of food areas, floors must be constructed so that they are: capable of being cleaned effectively, durable, non-slip. A gentle slope (1:50) to the drainage point is desirable as is a coved join between wall and floor. Many finishes are available.

- Terrazzo and granolithic (Grano) floors are very durable but smooth terrazzo (a marble-like appearance) is likely to be too slippery for food area use.
- Grano floors use crushed granite laid in a concrete base and are resistant to most substances found in food areas.
- Tiled surfaces are adequate if the joins between tiles are properly grouted.
- Epoxy resin mixed with aggregate is an effective but expensive floor finish.
- Welded heavy-duty vinyl sheeting is a cheap but often not long-lasting solution.

Flora (body) A term used to describe the population of micro-organisms living in and on a human or animal body. It includes those found both on the outer surface (skin, hair, fur, feathers) and those which inhabit the 'inside' of the animal (nose, mouth, throat, gastrointestinal and urinary tracts). The term includes both resident and transient groups. (The word 'flora' is used due to the original mis-classification of bacteria as plants.)

Flour mite (*Acarus siro/Tyroglyphus farinae*)

- A common pest of stored cereal products and domestic food stores.

They are less than 0.5 mm in length and have a colourless body with reddish/yellowish legs.
- They do not breed in conditions of low humidity (less than 75% RH) or in foodstuffs with water contents of less than about 13%.
- The most effective control measure is to ventilate food stores and keep cereal products dry.
- Mites are generally light-shy and will move slowly from bright light, resembling a moving dust when present in large numbers.
- There is a characteristic smell of mint from crushed mite bodies.
- Another mite (*Cheyletus eruditus*) is parasitic on the flour mite and mixed infestations are often encountered.

Flow diversion valve An automatic device used in pasteurisation and sterilisation plant (e.g. for milk/liquid egg) to return under-heated products to the start of the process. The valve is required by law in the UK and diverts liquid if its temperature has not reached the required levels.

Fluidised bed freezing A method of quick-freezing in which small items such as peas are frozen while suspended in currents of cold air passing through holes in a perforated plate. The bed may be inclined so that food travels along by gravity. Food is individually frozen with the advantage to the consumer that it may be 'poured' from the pack in any desired quantity.

Flukes Parasites of the bile duct, blood and lungs of various animals including humans. *See*: *Fasciola hepatica*, *Schistosoma*.

Fluoroacetamide A highly toxic substance used as a rodenticide in sewers. Unsuitable for use in food areas.

Flypapers As a control measure for wandering flies these may prove effective but must not be located near open food. They are also unsightly and may cause adverse reaction from customers or others. They are often used near to refuse, in stores or in and passageways between rooms.

Fo value The number of minutes heating at 121 °C needed to kill a particular micro-organism. Examples include:
- *Clostridium botulinum*, Fo = 3
- *Bacillus stearothermophilus*, Fo = 6.

This information is of use in canning and other forms of heat preservation. The measurement depends on the slope of the thermal death time curve.

Fomites Inanimate objects such as clothing, bedding, cutlery and crockery, which may distribute pathogens.

Food In Britain this is legally defined and includes everything normally thought of as food. It also includes drinks, additives, ingredients, water used in food production, slimming aids, dietary supplements and chewing gum. *See*: Appendix 1: Food Law.

Food Act 1984 A major piece of legislation now greatly amended by the Food Safety Act 1990. *See*: Appendix 1: Food Law.

Food-borne disease/Food-borne infection Typhoid and paratyphoid fevers, brucellosis, listeriosis, dysentery, hepatitis, cryptosporidiosis, giardiasis, cholera, campylobacter enteritis and tuberculosis are all examples of diseases which may be carried by food.

Characteristics

- Small numbers of organisms which use the food as a vehicle only and may not multiply in it.
- Symptoms produced are often different to food poisoning although diarrhoea and vomiting sometimers occur.
- The organisms may enter the bloodstream and cause infection. Incubation periods are often much longer than for food poisoning.
- These diseases may also be spread by non-food items.

Food business Defined in English law by the Food Safety Act 1990 as 'commercial or non-commercial activities which deal with food for human consumption'. Since profit is not included in the definition, schools, hospitals, clubs and other non-profit making organisations are covered by the definition. All must therefore comply with food hygiene legislation.

Food-contact surfaces Work surfaces which come into contact with food. They should:

- not affect the food in any way
- be smooth and impervious to allow for proper cleaning
- be non-absorbent.

Food grade stainless steel and certain plastic materials are considered to be the most suitable. The use of wood is unsuitable for all but raw meat preparation although a genuinely effective alternative has yet to appear.

As a general rule, food-contact surfaces should be cleaned and disinfected after each use with raw or high-risk foods.

Food handler A somewhat self-explanatory term with many possible categories including food production workers, chefs and bar staff. The hygiene training of food handlers should be related to the risk they pose to food. For example, a person working in a hospital cook-chill production unit may need more knowledge of food hygiene than a person only employed to stack shelves in a supermarket. *See*: Training of food handlers.

Food hygiene The term hygiene is derived from the name of the Greek goddess of health, Hygeia. Food hygiene includes:

- all aspects of health related to the consumption of food
- the prevention of food poisoning
- legislation
- design of equipment and premises
- personal hygiene
- staff training
- pest control
- food preparation practises
- cleaning systems
- storage of foodstuffs
- refrigeration and heating (temperature control)
- refuse and waste handling
- monitoring
- quality control.

Food Hygiene (Docks, Carriers, etc) Regs 1960/1962 Similar to the Food Hygiene (General) Regs 1970/1990 and designed for specific applications in docks, warehouses, cold stores, carriers' premises and certain vessels.

Food Hygiene (General) Regs 1970 and (Amendment) Regs 1990 The major food hygiene legislation in English law dealing with definitions, equipment, premises and food handlers. The important issues of high-risk foods and temperature control are also covered. *See*: Appendix 1: Food Law.

Food Hygiene (Market Stalls and Delivery Vehicle) Regs 1966 Similar to the Food Hygiene (General) Regulations but targeted at markets and delivery vehicles. *See*: Appendix 1: Food Law.

Food Labelling Regulations, 1984/1990 A full entry for these is to be found in Appendix 1: Food Law.

Food Law Modern food law has developed over a period well in excess of 100 years and is currently enshrined in the Food Safety Act 1990. Previous major Acts (not with the same title) are dated 1984, 1955 and

1938. Some of the first food laws in England during the 1800s concerned the adulteration of foods such as bread (with chalk and alum) and milk (with water), although religious food laws date back several thousand years. Periodically the law is modified because of:

- changes in the food chain
- technological changes
- the appearance of new trends in food poisoning
- pressure from various groups.

At present food hygiene issues are under EC consideration for the purpose of harmonisation. *See*: Appendix 1: Food Law.

Food (open) Unwrapped or unpackaged food. It should be kept covered or effectively screened during delivery or while exposed for sale.

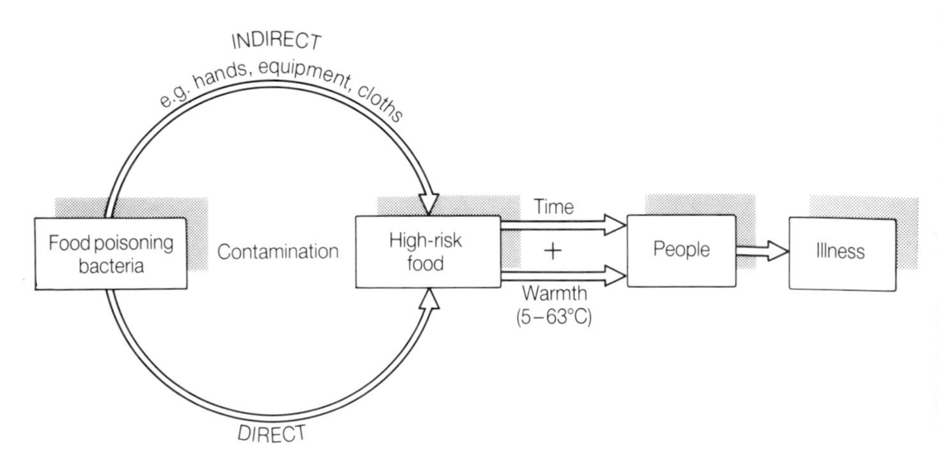

Food poisoning chain

Food poisoning A general term loosely applied to diseases which result in a disturbance of the gastrointestinal tract, normally affecting the stomach and intestines.

Food poisoning is mainly caused by:

- bacteria or their toxins
- viruses
- metals
- chemicals
- poisonous foods.

Symptoms of food poisoning include:

- vomiting
- diarrhoea
- fever
- pain.

Severe food poisoning can be fatal.

The causes of food poisoning, such as bacteria, do not affect the taste, smell or appearance of the food concerned.Those most at risk include the very young, the elderly, people who are ill and women during pregnancy. The most commonly reported food poisoning bacteria in many countries is *Salmonella*. In Britain, almost as many cases of *Campylobacter*, a food-borne disease occur annually. *See*: High-risk foods, Incidence of food poisoning, Temperature control.

Food premises All parts of the food business, including delivery, storage, preparation, cooking, food service, washing up, toilets, refuse disposal and any other relevant areas are included as part of the food premises.

General requirements include:

- adequate drainage
- water supply
- sanitary conveniences
- wash-hand basins (with related items)
- first aid materials
- accomodation for clothing
- separate facilities for washing food and equipment
- adequate lighting and ventilation
- cleanliness
- good repair
- facilities for keeping certain foods at specified temperatures.

See: Appendix 1: Food Law, and individual entries for above items.

Food preservation Deterioration and then spoilage of food starts from the point of harvest or slaughter due to the actions of several groups of micro-organisms including bacteria, moulds and yeasts and also enzymes. The aim of the various preservation processes is to halt this decay and so make available food which would otherwise be lost.
Food may also be eaten out of season and surpluses may be kept for use in times of shortage. Some methods keep food in a condition closely

resembling fresh while other methods cause great changes to occur. The history of food preservation dates back several thousand years (e.g. sun-drying) and new techniques (e.g. irradiation) are constantly being developed. Examples of food preservation include AFD, additives, bottling, canning, cook-chill, cook-freeze, chemicals, chilling, dehydration, freezing, irradiation, osmotic methods (salting/sugar), pickling, sous vide and vacuum packing. *See*: Individual entries.

Food preservation (nutrient losses) Vitamins of the B group (particularly B1, thiamin) as well as vitamin C (ascorbic acid) are destroyed in varying amounts by heat in some preservation processes including canning and dehydration. The losses may be considerable; for example, 100% loss of vitamin C occurs in some canned peas. Other vitamins are affected to a lesser degree.

Food processing Alternative term for food preservation and also used to indicate that a foodstuff has originated from large-scale production. The same term is used in small scale cookery to describe mechanically prepared foods.

Food room Under the Food Hygiene (General) Regs 1970 this is defined as any room in a food business in which food is handled.

Food Safety Act 1990

- The main piece of food legislation in English law covering many aspects of food and food hygiene.
- The Act defines the legal meanings of many terms including 'food', 'sale' and 'food business'.
- It sets out offences including those of selling food 'not of the nature, substance or quality', and food 'injurious to health'.
- It deals with false labelling and description of foods as well as the wider offence of not complying with food safety requirements.
- The defence of 'due diligence' is contained in the Act.
- Penalties of up to £20 000 are possible for certain offences.
- The power of the EHO is extended in terms of improvement notices and prohibition orders.
- Many regulations are made or are in the process of being made under the Act, including registration of food premises, training of food handlers and the licensing of food irradiation facilities.

See: Appendix 1: Food Law.

Food spoilage Deterioration of the quality of food, not to be confused with food poisoning. The common food poisoning organisms such as

Salmonellae do not normally cause spoilage to occur. *See*: Bacteria, Enzymes, Food preservation, Moulds, Organoleptic assessment, Yeasts.

Food Standards Committees Advise Ministers on compositional standards in foods, for example on meat content in pies, allowable levels of additives and labelling of foodstuffs. A large number of regulations governing all types of food have been made in this way having taken into account the requirements of manufacturers, EHOs, the general public and other parties.

Forced convection oven A method of reheating (regenerating) quantities of chilled or frozen foods using a fan-assisted oven in which the same temperature is achieved throughout the oven. Often used in conjunction with cook-chill and cook-freeze systems.

Foreign bodies *See*: Physical contamination.

Formic acid (including calcium and sodium formates) preservatives not permitted in Britain.

Freeze drying *See*: Accelerated Freeze Drying.

Freezer burn The unprotected surface of frozen food dries out (desiccates) giving a whitish, spongy appearance. The problem is often found in meat and is the result of ice crystals turning into water vapour by sublimation. Effective packaging is the answer to this problem.

Freezers (temperature of)
- Commercial and domestic deep freeze equipment operates in the temperature range −18 to −30 °C, the lower temperatures being found in deep freeze stores.
- Frozen food has a shelf-life of up to a year since low temperatures greatly reduce the activity of spoilage organisms.
- Some bacteria in frozen food are killed but many survive, become dormant, and will start to multiply once thawed.
- Temperature monitoring is important since at around −10 °C spoilage organisms start to grow.
- Deliveries of frozen food must be checked to ensure that food is still frozen; it has become common practice to reject deliveries above −18 °C.

See: Danger zone, Star markings, Thermometers.

Freezers (use of)
- Food should be placed in the freezer immediately on delivery.
- A system of stock rotation must operate.

- Freezer space should be sufficient to prevent overloading. Display type freezers have a maximum load line, which should not be exceeded.
- The freezer should be located away from heat sources such as cooking or heating equipment and not exposed to direct sunlight.
- Opening of doors on upright models affects inside temperatures.
- Damaged packaging of frozen food may lead to freezer burn.
- Generally, most meats, fruit and vegetables may be kept for 9–12 months and other foods for 3–6 months, following manufacturers' instructions.
- Deep freezers are not designed to produce good quality frozen raw food but may successfully freeze cooked items.
- Re-freezing of thawed food even if this has been cooked is not advisable, since the numbers of bacteria (bacterial load) increase through each freezing and thawing cycle.

Freezing (effect on bacteria) Significant numbers of food poisoning bacteria are not killed by freezing and spores remain virtually unaffected. Some organisms, principally those involved in food spoilage, are active well below 0 °C. *See*: Psychrotrophs, Thawing.

Fresh fish tapeworm (*Dibothriocephalus latum*) Parisitic on freshwater fish and salmon, it is also known to affect humans. The cysts in the fish appear as white growths.

Frozen food (production of) Low temperature preservation of foods has been performed for thousands of years although the familiar modern products are a recent development. Quick freezing causes ice to form within the cells of the food as small crystals. With slow freezing, water is withdrawn from cells, forms larger ice crystals and is lost as drip. Technically, freezing stops bacterial growth because of a reduction in available water (a_w for ice is 0.85) but has little effect on spores and toxins. It is achieved by a variety of methods. *See*: Available water, Birdseye (Clarence), Blanching, Blast freezer, Cook-chill, Cook-freeze, Cryogenic freezing, Fluidised bed freezing, Plate freezer, Zone of maximum ice crystal formation.

Frozen food (thawing and cooking of)

- Small items of food may be cooked direct from frozen.
- Large items such as joints of meat and poultry must first be thawed otherwise the heat of cooking will be wasted in melting ice and the final temperature will be too low.
- Thawing in a refrigerator takes longer than at room temperature but the

numbers of bacteria present after thawing will be less; if cooking is adequate then most bacteria are killed in any case.

- Ideally, a thawing cabinet or cool room (10–15 °C) away from other food should be used.
- Microwave and other electrical methods may be satisfactory as long as thawing is even and localised hot and cold spots do not develop.
- Any liquid from thawed food may contain large numbers of bacteria and should not be allowed to contact other food.
- Any equipment used in thawing must be disinfected before further use.
- As with other cooking, the final centre temperature should be at least 70 °C.

See: Cross-contamination, Thermometers.

Fruit Generally there are few food hygiene problems with fruits as they are low protein and sometimes high-acid foods. Deterioration during storage is prevented by storing unwrapped in cool conditions and ensuring adequate air circulation. Occasional outbreaks of food poisoning have resulted when acidic fruit juice has dissolved metals into foods.

Fruit fly (*Drosophila*) Several species including *Drosophila repleta*, *D. funebris* and *D. melanogaster* feed on fruits, vegetables, wine, vinegar and faecal matter. Used unwashed milk bottles have provided breeding grounds for these flies.

Fruit taint Fatty foods, for example meat and butter, readily absorb odours from oranges and other fruits. Correct packaging and storage conditions prevent this.

Frying A cooking method involving heat transfer through hot oil or fat rapidly achieving high surface temperatures in foods. There is a risk that the centre temperature may be inadequate to destroy any pathogens present.

Fugu fish A delicacy in Japan but unfortunately some internal organs contain a highly-toxic substance. These organs must be removed with surgical precision prior to consumption but occasional fatalities occur. The problem is compounded since the level of toxin appears to be at its highest at the time this fish is traditionally consumed.

Fumaric acid An organic acid used as a preservative in packet mixes and other foods. It has an antioxidant action.

Fumigation Buildings or parts of building may be filled with toxic substances such as methyl bromide or hydrogen cyanide to eliminate pests such as cockroaches. The procedure is hazardous, requiring expert operators and safe systems of work. The Environmental Health Department and Health and Safety Executive need to be informed prior to treatment. Fumigation often does not kill insect eggs and reinfestations are sometimes reported some months after when these have hatched.

Fungi One of the five kingdoms of living organisms. They are a very large group and consist of forms which vary from the unicellular (the yeasts), through the multicellular moulds and mildews, to the large multicellular types such as the mushrooms. Such large types look like plants, and indeed the fungi were originally mis-classified as such. However, the fungi are incapable of photosynthesis, a fundamental property of plants. Many fungi are edible, but there are some poisonous forms such as the Death Cap mushroom.

- The majority of fungi are saprophytes in soil or water, where they decompose plant material.
- They acquire their food by absorbing solutions of organic material from their surroundings, and so are often responsible for decomposing matter.
- They are responsible for much spoilage of foods.
- Many fungi are pathogenic to plants (such as the rusts and mildews), causing economic losses; a few forms are pathogenic to animals, including humans.

See: Aflatoxin, Food spoilage, Mycotoxins.

Fungicide Substance capable of killing fungi, an example being benzoic acid.

Furunculosis

1) A highly infectious and contagious disease of freshwater fish caused by the bacterium *Bacillus salmonicida*. Affected fish have ulcerating boils or furuncles under the skin and are unfit for human consumption.
2) The term also describes an outbreak of boils in humans, caused by *Staphylococci*.

See: Lesion (skin), Personal Hygiene, Pus.

Galvanised steel Mild steel coated with a thin layer of zinc is protected from corrosion and termed 'galvanised'. Zinc is dissolved by acid foods and has caused food poisoning, an example being the fermentation of cider apples in a so-called 'tin' bath.

Game Certain wild animals, mainly birds (such as pheasant, grouse and partridge) but also including hares, which are not eaten fresh but traditionally are 'hung' for some days to develop flavour and tenderise the flesh. This is achieved by the action of bacteria and enzymes, the final state being termed 'high'. Other meats in this condition would probably be considered unfit to eat. There is a legally defined season for hunting game, which may only be sold by licensed dealers.

Gannymede system A heated tray conveyor belt system of meal assembly often used in institutional catering such as hospitals.

Garchey system A method of refuse disposal in which waste water is used to carry rubbish by gravity from high-rise buildings.

Gas flushing A technique for modifying the atmosphere inside a pack by closing in a current of gas such as carbon dioxide or nitrogen. It is used to control the growth of aerobic spoilage organisms.

Gastronorm A modular oven for roasting and baking used in commercial kitchens.

Gastroenteritis A general term applied to diseases which cause inflammation of the stomach and intestines. Symptoms usually include one or more of the following: nausea, vomiting, abdominal pain and diarrhoea.

Gastrointestinal tract Another name for the digestive system, which essentially consists of the stomach and intestines.

Gelatin A high-risk meat product which has been involved in food poisoning due to contamination, followed by storage at too low a temperature. Gelatin melts at less than 40 °C and if it is kept around this temperature, conditions are ideal for the growth of any bacteria present. When molten it should be kept at a temperature of above 63 °C, and after use (e.g. as a glaze) it should be cooled to gelling point as rapidly as possible.

General outbreak (of food poisoning) An often widespread series of cases involving at least two people in different families.

Generation time The time taken for bacteria to double in number by binary fission. It depends upon all growth factors and is often quoted with particular reference to temperature. Many *Salmonella* species have a generation time of around 10 hours at 10 °C. *E. coli* has been known (given ideal conditions) to double in number in as little as 10 minutes. *See*: Bacterial growth curve, Danger zone.

Genus A classification name (plural: genera) given to groups of species that are very similar. The first letter is always written as a capital. For example: *Bacillus*, *Clostridium* and *Salmonella* are three genera of bacteria. *Aspergillus* and *Penicillium* are two genera of moulds. Together with the species name, it forms the scientific name of the organism.

Gerber test A laboratory method for determining the amount of fat in milk.

German cockroach (*Blattella germanica*) A brownish-yellow insect which can be up to 15 mm in length. It is also called the steam fly due to its preference for warm conditions. It is found in kitchens, bakeries and centrally heated buildings such as hospitals but it is also prevalent in many hot countries. The adult is active, agile and capable of rapid movement. It is also able to climb glossy surfaces. Any food may be infested, the insect is also gregarious and often found in large numbers. *See*: Cockroaches.

Germicide Any agent designed to kill 'germs', i.e. micro-organisms/ microbes. *See*: Bactericide, Fungicide, Virucide.

Giardia lamblia A flagellated enteric protozoan which has been shown to be an important cause of endemic and epidemic diarrhoea. The number of cases of the disease (giardiasis) in Britain has shown a five-fold increase from 1970 to 1990. The life cycle of the organism has two stages, a free-living stage and the cyst. Infection occurs after ingesting a small number of the cysts. The free-living forms are destroyed by the acid contents of the stomach. About half of the people swallowing the cysts develop diarrhoea. Other symptoms include abdominal cramps, bloating, flatulence and nausea. Vomiting and fever are less common. Symptoms persist for several weeks, with cysts being released in the faeces. Infection is acquired from contaminated water or direct contact with faeces-contaminated individuals or objects. *Giardia* is sensitive to chlorination, but the cysts survive standard concentrations of chlorine used by most waste water treatment plants. Higher levels of 4 to 6 mg/dm^3 are required.

Glanders A notifiable disease of horses, affecting mucous membranes of the nostrils, which may fatally affect humans. The causative organism is the Gram-positive bacterium *Bacillus mallei*. The disease is also called farcy.

Glass (contamination by) A physical contaminant of foodstuffs normally resulting in court action since affected food is clearly injurious to health. In food premises, glass must be carefully controlled. Examples include automated scanning of foodstuffs to detect glass fragments.

'non-shatter' lighting, wired glass in doors and windows and the banning of glass items from production areas.

Glass washing Large quantities of glassware, for example in restaurants or the licensed trade, are best washed by a specifically designed machine with a final polish, if required, with disposable paper towels. *See*: Cleaning, Washing up.

Gloves Disposable gloves may be worn by food handlers to protect against staphylococcal contamination. Gloves can, however, become a vehicle for cross-contamination. Food-use gloves are less likley to cause physical contamination if brightly coloured (blue).

Goats *See*: *Brucella abortus*, Meat, Milk.

Goitrins Cyanide-containing compounds present in the cabbage family (*Brassica*). They may aggravate thyroid problems by interfering with the uptake of iodine which is used by the thyroid gland to manufacture the hormone thyroxine.

Golden spider beetle (*Niptus hololeucus*) A widespread pest in many countries; the adult, less than 10 mm long, is covered in golden hairs and resembles a spider. Cereal products and other foods are affected and the beetle may also feed on textiles.

Good housekeeping A self-explanatory term often translated as, 'A place for everything and everything in its place'. This should be part of the general cleaning and maintenance operations of all food businesses and has several benefits including discouraging pests, increasing safety and promoting a positive image.

Grain beetle or Grain weevil (***Sitophilus granarius***) A small brown weevil, up to 5 mm long, and having a characteristic snout, very widely occuring and contaminating many stored cereals and cereal products.

Gram stain A laboratory procedure used to dye bacterial cells. It is used as an aid to identification. The process involves applying two stains (or dyes) one after the other. Due to differences in the composition of their cell walls, all bacteria retain one or other of the two stains. The test is easy to perform, taking just a few minutes. Depending on which of the two stains is retained, bacteria are classified as either Gram-positive or Gram-negative. Knowing how pathogenic bacteria react to the Gram stain is of value in determining a course of action in cases of infection; e.g. antibiotics vary in their effectiveness against Gram-positive and Gram-negative organisms, and all sporing bacteria are Gram-positive.

Gravy Made from meat juices, so it is a high-risk food. Temperature control is therefore important; it must be kept refrigerated (1–4 °C) or hot (over 63 °C).

Grease trap In drainage systems, a device to trap grease and thus prevent it solidifying or congealing and causing blockages. These should be located outside food rooms and cleaned at regular intervals.

Green mould rot A type of food spoilage. *See*: *Cladosporium*.

Growth curve (bacterial) *See*: Bacterial growth curve.

Growth requirements (bacterial) Micro-organisms which cause food poisoning must be able to grow in food. In order to grow they need the following:

1) **Food** Bacteria will grow in most foods, but they will grow best in foods rich in protein, e.g. meat and meat products, fish, poultry, eggs and egg products, milk and milk products.
   See: High-risk foods.
2) **Warmth** Human pathogens grow best at body temperature (37 °C), but they will grow and reproduce both above and below this temperature. The effective range is 5 to 63 °C.
   See: Danger Zone.
3) **Moisture** Like all living organisms, bacteria need moisture in order to survive. If a food is dehydrated bacteria are unable to grow or reproduce. However, as soon as such foods are rehydrated, bacteria will start to grow and reproduce. Hence rehydrated foods must be treated in exactly the same way as fresh foods.
   See: Dry goods stores, food preservation.
4) **pH** Most bacteria prefer a pH of around 7.0 (neutral). However, some bacteria are able to grow down to around pH 4 (very acid).
   See: Acid foods (classification), Pickling.
5) **Gaseous environment** Some bacteria need oxygen to survive (aerobes), but others cannot grow in its presence (anaerobes). Some can grow under both conditions (facultative organisms).
6) **Time** Given ideal conditions, bacterial cells can grow and reproduce in as little as ten minutes. This means that extremely large numbers of bacteria may be produced in a short period of time.
   See: Mean Generation Time.

Growth requirements (fungal)

1) **Food** Require less nitrogen for growth compared to bacteria. They are nutritionally more efficient, and so can grow on such

things as leather. In general, sugars are best for yeasts; moulds grow well on a wide range of foods, both simple and complex.

2) **Temperature** Generally less heat tolerant than bacteria, but they do grow better at lower temperatures. Optimum temperature is from 25 to 30 °C, maximum temperature for growth is in the range 35 to 47 °C. Some can grow at 0 °C or less.
3) **Moisture** They are more able to grow on foods with too little moisture to support bacterial growth. Yeasts require more moisture than moulds. Most fungi are more resistant to high osmotic pressure than bacteria, and so readily grow in jams, for example.
4) **pH** Fungi usually grow better in an acid pH, yeasts in particular around pH 4 to 5.
5) **Gaseous environment** All moulds growing in foods require oxygen (aerobes). Yeasts can grow in both the presence and absence of oxygen (facultative organisms). If oxygen is present, carbohydrates are broken down to carbon dioxide and water; if it is absent, carbon dioxide and alcohols are produced. This fact forms the basis of the baking and brewing industries respectively.

Gut An alternative name for the intestines.

Habits (personal) Unconscious actions involving contact between fingers and other parts of the body, for example the nose, the unguarded cough or sneeze, licking the fingers to pick up paper for wrapping food, etc. Such actions may cause bacteria or other contaminants to be passed onto foods.

HACCP *See*: Hazard Analysis Critical Control Points.

Haemagglutin (Lectin) A toxic substance found in various pulses including red kidney and haricot beans.

Hair Stray hairs may become physical contaminants of food and may also be contaminated with *Staphylococcus aureus* from the scalp. A total hair covering, using a hairnet where necessary, should be put on before entry into foodrooms.

Haloduric 'Salt-tolerant'. Certain staphylococci are haloduric and so will live (but not multiply) in salty foods such as ham and also on salted wooden chopping boards. It is not a property of most food poisoning bacteria.

Halophile 'Salt-liking'. Some organisms are able to multiply in high concentrations of salt, e.g. *Halobacterium* (a_w 0.75) which causes discolourations (pink spot) on salted fish.

Ham

- A cooked, cured pork product involved in many cases of food poisoning due to inadequate temperature control after cross-contamination.
- Ham is sometimes contaminated by hand/machine-contact during slicing or food preparation.
- Prior to consumption ham must be refrigerated.
- Contact between ham and raw meat should be avoided.
- It is particularly prone to contamination by staphylococci from food handlers, due to the bacterium being able to tolerate the high salt content of the product.

All other cold cooked meats share these risks.

Hand-contact surfaces A self-explanatory term; such surfaces (e.g. work surfaces and door handles) require disinfection to ensure that any bacteria transferred to them by the hands are not passed on to food or other items. *See*: Cleaning schedules, Food-contact surfaces.

Handkerchiefs Become contaminated with bacteria such as staphylococci, and after use the hands should be thoroughly washed. It is questionable if the use of these should be allowed in food rooms since the user may not be in a fit condition to handle food if suffering from a heavy cold. Generally, single-use disposable tissues are preferred.

Handles *See*: Hand-contact surfaces.

Hands

- The hands of a food handler are probably the most dangerous part of their body.
- Hands may become covered with bacteria after handling raw foods such as meat and poultry.
- They are often the vehicle in cross-contamination as well as being reservoirs of bacteria such as *Staphylococcus aureus*.
- Fingernails should be kept short and not bitten.
- Nail varnish (a possible physical contaminant) should not be worn.
- Perfumed hand creams may contaminate foods.
- Cuts must be covered with a suitable waterproof dressing.
- Jewellery is restricted to a plain wedding ring which cannot collect food scraps and harbour bacteria.
- Infections such as septic cuts or whitlows must be reported and contact

with food or related equipment avoided until medical clearance is obtained.

- Frequent handwashing is vital, for example before starting to work with food and between handling different foods such as raw and cooked.

Hand-washing Food handlers are required to wash their hands on many occasions including:

- before entering the food room and starting work
- after using the toilet
- after smoking
- when refuse or waste has been handled
- after leaving the food room for a break
- after handling any raw food
- after any other occasion that they have become contaminated.

Soap, hot water, a separate basin, suitable drying facilities and a nailbrush must be provided and a notice requesting hand-washing must be clearly visible in toilet areas.

Harbourage A term used to mean shelter for pests. Examples of harbourage for rats and mice include cavity walls, roller boxes for shutters, spaces behind and within equipment, suspended ceilings and badly stored foodstuffs.

Hard cheese Cheeses such as Cheddar, Parmesan and Emmenthal, produced by a process including pressing, in which the moisture content is 30–40%. Storage of these is at 10 °C during ripening and 5–10 °C afterwards. Any cooked products containing cheese must be kept out of the danger zone of temperature. Occasional outbreaks of staphylococcal food poisoning have been reported with cheese made from unpasteurised milk.

Hardness of water Caused by various dissolved substances such as calcium and magnesium salts. Hard water:

- will not allow water to lather with soap
- causes scum formation
- results in deposits on heating equipment affecting safety and efficiency.

Hardness may be temporary (caused by hydrogen carbonates) which may be removed by boiling, or permanent (caused by sulphates) which must be chemically removed by water softeners such as Calgon™ or Permutit™.

Hard swell A can of food, the ends of which are permanently protruding due to gas production; also called a 'blown' can.

Hazard Analysis Critical Control Points (HACCP) A technique which involves:

1) identification of all potential hygiene problems throughout a food production process, and
2) developing strategies to deal with them.

The method is widely used in food manufacturing and cook-chill and pays particular attention to the following points:

- where bacteria and other contaminants may enter the process and contaminate food,
- temperature control,
- cross-contamination.

The whole process (from raw materials through to consumption) must be considered, including cleaning systems and monitoring of product quality. It is common to have an extremely detailed specification for each product to include all ingredients, processes and uses.

Head coverings *See*: Hair.

Health and Safety at Work, etc Act 1974 A major piece of legislation dealing with all aspects of safety at work. It applies to employers, employees and members of the public. Employers have major responsibilities and must ensure that workplaces are safe and that staff receive adequate training and information about hazards at work. The Act and its many regulations is jointly administered by the Health and Safety Executive and Local Authorities depending on the type of workplace.

Health and Safety Commission and Executive (HSC and HSE)

1) The HSC is an independent body established under the 1974 Act (see previous entry) consisting of representatives from employers, trades unions and professional bodies, for which the Secretary of State answers to Parliament. The HSC has responsibilities including research, administering health and safety legislation, advising, preparing new legislation for consideration and investigating safety issues such as accident prevention.
2) The HSE is essentially the enforcement agency of the HSC operating 'in the field'. HSE inspectors are divided into several specialist divisions (such as pollution) and work from regional centres. Generally HSE inspectors are involved with factories, construction sites and educational establishments. HSE also publishes a large amount of information relating to all aspects of safety.

Health Inspectors Popular term used for Environmental Health Officer and derived from the pre-1974 title of 'Public Health Inspector'.

Healthy carriers Persons who excrete food poisoning or other disease-causing organisms without ever having shown symptoms. They may pass a disease unknowingly through food to other people. *See*: Carriers.

Heat treatment of milk and other foods Milk may be made safe by pasteurisation, sterilisation or ultra heat treatment (UHT), all of which destroy pathogenic organisms. Canning is the most common heat treatment preservation process for other foods although cooking could also be considered as such. *See*: Milk.

Heavy metals *See*: Metals.

Hepatitis A (HAV) Hepatitis is an infectious disease of the liver and is caused by a group of viruses. Hepatitis type A virus (HAV) causes epidemic or infectious hepatitis because of its mode of transmission. It is usually spread by the faecal-oral route, but not infrequently outbreaks occur through the transmission of the virus in water, food and milk. Foods involved have included salads, raw and steamed shellfish and bakery products. Proper hand washing before food preparation and serving will effectively prevent such outbreaks.

Hepatitis B (HBV) A viral disease spread through mucous membranes and carried in body fluids such as blood. It is only very rarely known to be food-borne.

Herbivore An animal that feeds exclusively on plants. Examples include cattle and sheep.

Herbs and spices The flavour of some of these may mask deterioration in foods. Cloves and mustard are known to be bactericidal and other herbs have medicinal uses. Some may be irradiated.

Herring Fish often smoked and cured to make various products including kippers.

Hexamine (Hexamethylenetetramine, E239) Fungicide used on some fish and cheese.

Hide beetle (*Dermestida, Dermestes* and *Anthrenus* species) Small, oval/round beetles (2 to 5 mm in size) with patterned bodies. They infest dry protein material including leather goods, biscuits, dead animals and birds, mummified bodies and occasionally food stores. The larvae can cause considerable damage by boring.

'High' *See*: Game

High-risk foods Foods which support the growth of bacteria, particularly pathogens, when conditions are favourable and which are intended for consumption without treatment which would destroy such organisms. These foods commonly contain protein and include:

- cooked meats, fish and eggs (and substitutes for any of these) and any product containing them (such as pies);
- cooked cheese, cereals (e.g.. rice), pulses and vegetables;
- soft cheeses ripened with moulds or micro-organisms;
- smoked or cured fish;
- smoked or cured meat which has been sliced;
- desserts containing milk, with pH 4.5 or more;
- prepared vegetable salads including those with fruit;
- sandwiches;
- cream cakes.

All of these foods should be kept under strict temperature control, generally out of the danger zone (5–63 °C). There are exemptions under English law to allow time for sale and service of some foods. *See*: Appendix 1: Food Law.

Home catering *See*: Domestic premises.

Homogenised Literally, the same composition throughout; used to describe milk where the cream has been converted into small globules which remain in suspension and do not separate into a top layer. The process involves forcing milk through a fine nozzle under pressure and is used before sterilisation and UHT processes.

Hot air hand driers A modern and effective method of hand drying eliminating the risk of person to person contamination. A disadvantage is their slowness which may discourage their use. They are best used with a backup system such as disposable paper towels in case of failure.

Hot cabinets (Hot cupboards) Equipment consisting of a heated metal cabinet with shelves for keeping food hot while awaiting service. Food in these should be at above 63 °C and the cabinet temperature over 80 °C. Such cabinets are not suitable for reheating food, their use being to hold the temperature of high-risk foods above the danger zone. For plate warming a temperature of 60 °C is satisfactory. *See*: Bain Marie.

Hot food High-risk foods should be kept above 63 °C to prevent the growth of food poisoning bacteria or the germination of spores. This temperature is lower than the service temperature for many foods, this being above 70 °C. *See*: Appendix 1: Food Law, Thermometers.

Hot water A legal requirement in food rooms for washing hands unless no open food is handled. *See*: Appendix 1: Food Law.

HSC and HSE *See*: Health and Safety Commission and Executive.

HTST 'High Temperature Short Time'. A method of pasteurisation often used for milk in which 72 °C is held for at least 15 seconds follwed by immediate cooling to 10 °C.

Humidity (Relative Humidity, RH) The amount of water vapour present in air at any one time, expressed as %RH. 100% humidity indicates that the air is saturated and cannot hold any more water vapour. Excess humidity (over RH 70%) is uncomfortable to work in since the body cannot lose heat by sweating, and it may cause condensation problems affecting buildings and stored foods. For dry food storage rooms, RH range of 30–70% is satisfactory (60% is ideal), whereas raw meat and most fruit and vegetables need RH of 90%. Humidity is measured with a hygrometer.

Humectant Substance which absorbs water from the atmosphere and prevents food from drying out, for example glycerol (glycerine) in icing.

Hung *See*: Game.

Hydatid disease/cysts The tapeworm *Echinococcus granulosus* lives in the gut of dogs, cats, foxes and other carnivores. It is a small tapeworm (1 cm long) whose eggs are eaten by an intermediate host including humans, cattle, pigs and sheep. The eggs produce cysts in many organs including the liver and brain. Direct contact with dogs or through food contaminated with dog faeces (such as uncooked vegetables) have been shown to be common ways for humans to contract this disease.

Hydrogenation The process of saturating some double bonds in unsaturated oils by passing hydrogen gas through the heated oil under pressure in the presence of powdered nickel as a catalyst. Many margarines are made in this way.

Hydrogen peroxide A substance with a bactericidal effect increasingly used in cleaning agents such as disinfectants. It is biodegradable, producing oxygen and water but is not as powerful as hypochlorites.

Hydrogen sulphide A gas with an offensive odour produced during the decomposition of eggs and during the cooking of cabbage and similar vegetables. *See*: Sulphiding.

Hydrogen swell Internal corrosion of cans resulting in a gas build up which produces 'blowing'. It mainly affects canned fruits containing acids which attack the tinplate and react with the mild steel of the can.

Hydrometer A flotation device for measuring the density (specific gravity) of liquids including wines, beers, fruit juices, sugar syrups and milk with the object of determining some element of composition, for example the alcohol content of wine, and solids in milk.

Hydrophilic Literally 'water-liking', a term used in detergency.

Hydroxybiphenyl, E231 *See*: Biphenyl.

Hygrometer *See*: Humidity.

Hypha A single slender filament composed of a number of cells, found in the fungi. *See*: Moulds, Mycelium.

Hypochlorite

- Salts of hypochlorous acid containing the radical $-Cl^{-}$ and used as disinfectants due to the action of chlorine released from them.
- Chlorine is bactericidal and is effective against some spores.
- At a concentration of over 100 ppm a fresh sodium hypochlorite solution will disinfect an aesthetically clean surface in 2 to 3 minutes.
- If the surface is dirty then disinfection may not be achieved unless the strength of the solution is increased; there is a limit to this since the chemical affects both the skin and metals.
- Solutions are often supplied at a concentration of 10% and require dilution before use, with rinsing after use to avoid taint.
- Strong hypochlorite solutions may damage the skin, eyes and fabrics and should be carefully handled.
- If mixed with other cleaning agents such as scouring powders, chlorine gas may be released, the inhalation of which could be fatal.
- Hypochlorites are anionic and should not be mixed with cationic substances.
- There is concern over the biodegradability of hypochlorites and related environmental effects, but they have wide use domestically and commercially, being readily available, effective and cheap.

See: COSHH, Contact time.

Ice Has been involved in cases of food poisoning (including salmonellosis) from ice cubes in drinks. Ice-making machines should be well maintained to avoid problems. Ice is included within the legal definition of food. *See*: Growth requirements (bacterial), Psychrophiles.

Ice-cream

- A high-risk product whose production and sale is closely controlled by Ice-cream (Heat Treatment) Regs 1959–63.
- Ingredients may include milk, eggs, sugar, thickening agents and water, many products having a large percentage of air ('overrun').
- Pasteurisation and sterilisation times and temperatures are legally defined with cooling to 7 °C within 90 minutes.
- The final product may be stored at −18 °C but special ice-cream refrigerators operating at −2 °C are also used for short-term storage (one week); the texture is best at around this temperature.
- If ice-cream thaws it must not be refrozen.
- Scoops for portioning ice-cream from bulk should be washed and disinfected between each service.
- With soft ice-cream it is vital to clean the machine regularly according to manufacturers' instructions.
- The most up-to-date soft ice-cream machines are self-pasteurising.

Ichneumon flies Minute two-winged parasitic insects which attack pests of stored products by laying eggs into larvae or caterpillars. Their presence is an indication that pest infestation is at a high level.

IEHO Institution of Environmental Health Officers. The professional body representing EHOs, based in London, England. It has responsibilities regarding training, issue of guidance notes and Codes of Practice, reporting to Government and many others.

Ileum The last part of the small intestine.

Illness in food handlers

- Infectious diseases such as food poisoning, typhoid fever, gastroenteritis and others must be reported to the local authority.
- A decision is made as to whether or not the food handler may continue working.
- Faecal samples may be taken to identify the cause of the disease; as an example, if a food handler proves to be *Salmonella* positive they must provide negative samples on three consecutive days before being allowed to work with food *See*: Exclusion of food handlers.

Imitation cream Subject to the same risks as cream.

Immersion freezing *See*: Cryogenic freezing.

Imported foods Regulations governing these are found in many countries including all parts of the UK. Port Health Authorities employ EHOs whose duties include food inspection and the examination of health certificates required for some imports.

Impression plate A technique used to inoculate an agar plate by pressing the material to be tested, e.g. a cloth or part of an overall, directly onto the surface of an agar plate.

Improvement notice

- Under the Food Safety Act 1990 and other laws, a formal notice stating works to be done to bring premises or practises up to standard may be served on the proprietor of a food business.
- The notice is normally served after an inspection by an EHO or other authorised officer.
- The notice is subject to appeal and at least 14 days must be given to do work required; extensions to this time may be applied for.
- If the notice is not complied with then prosecution or prohibition may result. *See*: Informal notice, Prohibition orders.

Incidence of food poisoning and food-borne disease In Great Britain there has been a general upward trend in reported cases of food poisoning and food borne-disease in the period 1970–91. Total reported cases have more than doubled to around 60 000 but the number of non-reported cases, most of which may be relatively minor, is likely to be far in excess of this figure. Various estimates range from 1–10 million per year. *See*: Food poisoning.

Incubation period The time between consumption of contaminated food and the appearance of any signs or symptoms. The precise time varies from hours to days, and depends on a number of factors, including the micro-organism involved, the number of infecting micro-organisms, and the resistance of the host. *See*: Onset period.

Incubatory carrier *See*: Carrier.

Indian-meal moth (*Plodia interpunctella*) A pest principally of dried fruits and nuts, prevalent in California, Australia and Mediterranean countries. It also attacks chocolate and cereals and is cold-hardy, being able to survive British winters. *See*: Moths.

Indicator organisms Organisms whose presence is tested for when assessing water safety. Such organisms must:

- be consistently present in substantial numbers in human intestinal wastes,

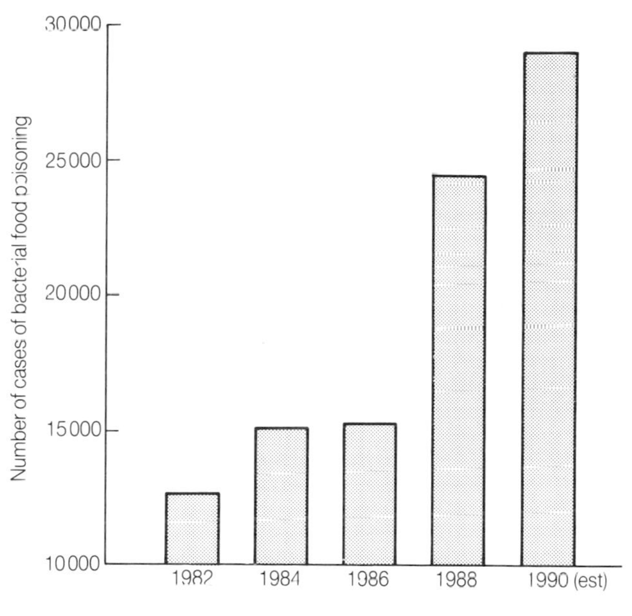

Incidence of food poisoning

- survive in the water for at least as long as pathogenic organisms,
- be detectable by simple tests.

The usual indicator organisms are the faecal coliform bacteria such as *E. coli*. *See*: Presumptive water (coliform) test.

Indictable offence Serious offences normally heard in a Crown Court.

Induction training Training for new employees covering principles of food hygiene and any other important matters such as safety. For food handlers a basic food hygiene course should be undertaken before working in any situation which could affect food safety. New employees may also require a period of supervision.

Informal notice
- A letter sent by an EHO following an inspection of a food premises stating defects and remedies.
- The notice has no legal status and is only used for non-urgent matters but if no action is taken then stronger measures such as an improvement notice may result.
- As the name suggests this notice is a form of 'friendly persuasion' and is

favoured by some as a means of getting problems dealt with without being too heavy-handed.
- Another school of thought suggests that it is a waste of time since it may be ignored because of its lack of legal weight.

Inoculate To transfer micro-organisms to, for example, a sterilised growth medium. After inoculation, the medium is placed in an incubator for one to two days to allow the micro-organisms to grow. *See*: Culture media, Incubator.

Inoculum A sample of micro-organism culture used to start a new culture.

Infection (food poisoning/food borne) An illness caused by the ingestion of food contaminated with living bacterial cells. There are two types of food infections:

1) Those in which the pathogens grow and multiply in the food (food poisoning).
2) Those in which the food does not support the growth of pathogens, but simply serves as a carrier for them (food-borne diseases).

Infections (water-borne) In the food industry, water has many functions, including:

1) as an ingredient
2) to wash ingredients
3) to heat/cool food
4) to clean.

In each case it is important that the water is free from pathogenic and spoilage organisms. A large number of pathogenic organisms may be carried by water.

Contamination by water

a) Direct, for example by drinking water. A variety of micro-organisms may be spread by this route, including viruses (such as the hepatitis A virus, the poliomyelitis viruses and the enteroviruses), bacteria (such as the organisms responsible for causing typhoid fever and cholera) and protozoa (such as the organism causing amoebic dysentery).
b) Indirect, for example through contamination of a food product, such as faecally-contaminated shellfish and plant crops. There is also the possibility of infection by a variety of worms such as cestodes, nematodes and trematodes in raw or under-cooked fish.

Infectious disease *See*: Communicable disease.

Infestation The presence of pests such as insects or rodents in or on food, premises or persons.

In-place cleaning *See*: CIP.

Insecticides Chemical formulations or natural substances used to kill insects. Examples include pyrethrum and diazonium compounds. Food room use must be strictly controlled by professional operators. *See*: Cockroaches, DDT, LD_{50}.

'Insect-o-cutor' A trade name for a type of electrical flying insect killer using ultra violet radiation to attract pests.

Insects A large class of invertebrates having head, thorax and abdomen, two antennae, three pairs of thoracic legs and normally two pairs of thoracic wings. Examples include flies, beetles and cockroaches.

Inspections of food premises

1) May be performed by authorised officers such as EHOs at any reasonable time, obstruction being an offence.
2) Regular and systematic walking tours of food premises using a prepared check list are a useful management tool to ensure that standards are being maintained.

Intestinal flora (organisms)

- Bacteria heavily populate most of the digestive system.
- The stomach and small intestine contain fairly small numbers, due in part to the acid nature of the gastric juice.
- The large intestine however contains enormous numbers of bacteria; it has been estimated that there are over 100 billion bacteria per gram of faeces.
- Up to 40 % of the weight of faeces consists of wet microbial cells, both living and non-living.
- In the adult lower bowel most of the organisms are commensals, but there will also be low numbers of potential pathogens, such as *Salmonella* species and *Clostridium perfringens*.
- The larger numbers of these organisms excreted by infected people or carriers may lead to an outbreak of disease, especially if their personal hygiene is poor.
- The non-pathogenic bacteria assist in the enzymic breakdown of foods and some produce useful vitamins, such as vitamin K.

Note: The word 'flora' is used due to the original mis-classification of the bacteria as plants.

Intoxication (food poisoning by) A food-borne illness caused by the consumption of food containing a chemical toxin (poison). The toxin may be a natural product found in certain plants or animals, or it may be a toxic substance excreted by a micro-organism such as a bacterium or fungus.

Invasiveness *See*: Aggressiveness.

IVS (Intervening ventilated space) Such a space, ventilated to the outside, must exist between a room containing a WC and another room.

Iodophors Iodine-containing sanitisers used principally by beverage manufacturers and dairies in CIP systems. They tend to be expensive but are effective in a wide range of conditions. Formulations are non-ionic and are buffered with phosphoric acid at pH 3–5.

Iron A common metal used for making cooking vessels and the mild steel plate of cans. When dissolved by acid foods, there is a risk of metallic food poisoning and also unpleasant tastes.

Irradiation

- A method of food preservation developed as a result of research during the 1960s.
- Radiation at levels around 10 KGy (gamma and X rays) or 10 MEv (electrons) from a radioactive source or generator will kill micro-organisms and insect pests.
- High doses (sterilising doses) of radiation cause severe changes to the taste of foods making them unacceptable to consumers and current practice is to use lower levels to reduce the total bacterial load.
- Low doses are also used to prevent potatoes, onions and other vegetables from sprouting.
- Many foods such as chicken, shellfish, strawberries and spices preserved by this method have been on sale worldwide for some years.

Main problems

- The process does not affect spores or toxins and is not a substitute for good food hygiene practises.
- Much concern has been expressed about the possibilities of induced irradiation in consumers, mutation of bacteria and the increased risk of survival of some pathogens since spoilage organisms are more easily destroyed.
- There is no simple reliable test to detect irradiated food.
- Consumer studies have shown that many people would be reluctant to eat irradiated food for various reasons including safety issues.

Producers of irradiated food in Britain must be licensed. Production of irradiated food is controlled by the Food (Control of Irradiation) Regs 1990 and the Food Labelling (Amendment) (Irradiated Foods) Regs 1990.

Jam Fruit product preserved by sugar and having a shelf-life of up to a year if unopened. Mould growth may cause spoilage. Refrigerated storage after opening is often recommended.

Jejunum Part of the small intestine.

Jewellery When worn by food handlers this creates several risks including:

1) trapping food particles upon which bacteria may breed,
2) the possibility of physical contamination if any part falls into food,
3) various safety hazards.

It is common practise for food handlers to be allowed to wear only a plain wedding ring.

Juvenile hormone A synthetic hormone used in insect control (Pharaoh's ants and cockroaches) which when eaten by the young insect prevents their development into adults. The hormone is species-specific and should not affect other life forms.

Kickplate A panel fixed to the bottom of a door to minimise damage. It may be made of a number of materials, such as plastic, but if metal is used it is more durable and it helps proof against rodent pests. This latter property is further enhanced if it makes a close fit on closing.

Kidney bean *See*: Red kidney beans.

Kipper A cured, dyed and smoked herring. *See*: Curing, Fish, Smoked foods.

Kitchen design *See*: Design of food premises.

Knacker's yard A slaughterhouse dealing with animals not intended for human consumption.

Koch, Robert (1843–1910) A German physician and microbiologist, whose pioneering studies of anthrax, tuberculosis and cholera showed them to be bacterial diseases. He is regarded by many as the first true bacteriologist.

Kosher Food prepared according to Jewish traditions.

Labelling It is an offence under English law to label, describe, advertise or present food in a way that is likely to mislead, based on the expectations of an ordinary person. Examples include:

- making a false claim (describing a cheap cut of meat as fillet steak or using a picture of a dairy cow on a food containing no milk),
- failing to declare information (such as the use of irradiated food).

All types of printed material including wrappers, circulars and menus, together with radio, TV and any other form of advertising are covered by these requirements. *See*: Appendix 1: Food Law.

Lactic acid An organic acid produced in foods by bacteria including *Lactobacillus* and *Streptococcus* species.

1) In cheese-making lactose is fermented to lactic acid, the resultant lowered pH being suitable for the clotting action of the enzyme rennin.
2) In slaughtering food animals, a rested animal which has a high level of glycogen in its muscles produces better quality meat. This is because glycogen is converted (post-slaughter) into lactic acid, which tenderises meat by breaking down the fibres of the proteins collagen and elastin. Meat with a low pH keeps longer since bacterial growth and enzyme action are inhibited.
3) In yoghurt and similar products lactic acid is produced from various sugars including lactose. The low pH causes coagulation of the milk protein casein and gives the product its characteristic taste. If the pH falls to less than 4.5, these products do not support the growth of common pathogens.

See: Dark-cutting meat.

Lactobacillus Gram-positive bacteria which ferment sugars including lactose to produce lactic acid. Most common is *L. bulgaricus* (*L. acidophilus*, *L. jugurti*, *L. lactis* and *L. helveticus* are also used in yoghurt making). Some species are sugar-tolerant and cause spoilage in sweet cured hams. In canned meat products a green appearance may result from the growth of these organisms. *See*: *Streptococcus* species.

Lag phase *See*: Bacterial growth curve.

Lacquer band In the treatment of infestations by insects such as cockroaches, a band of lacquer or varnish containing an insectide is painted around areas such as pipework. The insect walks over the lacquer and picks up the insecticide on its feet and body. The insecticide is ingested after grooming and/or carried back to the nest.

Larvae (insect) A developmental stage in the life cycle of some insects, between egg and pupa, for example the maggots of flies.

Latent heat Heat which must be added or removed to change the state of a substance, temperature being unchanged. For example a large amount of heat must be removed to convert water into ice at freezing point. Thawing of frozen foods often takes longer than would be expected since the melting of ice to form water requires the input of large amounts of heat.

Lavatories *See*: Sanitary conveniences.

Law *See*: Appendix 1: Food Law.

LC_{50} The concentration of a substance, such as an insecticide or rodenticide, which will kill 50% of those exposed to it. More often used is the unit LD_{50}.

LD_{50} The dose of a substance which will kill 50% of those exposed to it and thus is used as a measure of acute toxicity and as an indication of harmful effects to others at risk. The substance dieldrin, while very effective against insects, is also toxic to mammals having an LD_{50} of less than 100 mg/kg body weight in rats while pyrethrum, a common flykiller, has an LD_{50} of over 800 mg/kg and is thus 'safer' for mammals. *See*: LC_{50}.

Lead
- A heavy metal which may contaminate food from:
  water pipes, exhaust fumes, paints, some pottery glazes.
- Lead is a cumulative poison whose effects, including deterioration of the brain, may not be apparent for many years.
- The maximum 'safe' level in foods is around 1 ppm but an even lower figure may be more desirable.

Lectin *See*: Red kidney beans.

Legislation *See*: Appendix 1: Food Law.

Leptospira Bacterial organisms having a coiled appearance including those causing Weil's disease and Canicola fever, both being examples of leptospirosis. The organisms are carried by several animals including dogs and rats. Both diseases are water-borne.

Lesion (skin) An area of damaged skin, such as a graze, cut or burn. Such damaged tissues readily support the growth of bacteria and so there is a high risk of the wound becoming septic. It is compulsory under Food Law to cover such lesions with waterproof dressings. *See*: Appendix 1: Food Law, Personal hygiene, Pus, *Staphylococcus aureus*.

Leuconostoc Gram-positive food spoilage fermenting bacteria causing rot in vegetables. They are both salt and sugar tolerant. They are found in slimy sugar solutions, fermenting vegetables and in milk and dairy products. Some are able to spoil wine.

Leucosis (avian) A disease of the liver in poultry causing emaciation. Affected birds are unfit for human consumption.

Licensing and registration of food premises

- Certain types of food business, such as those making ice-cream, slaughterhouses and those dealing with certain aspects of milk production and sale, are registered with the local authority who may refuse to issue or renew the licence if operations are unsatisfactory.
- General registration of most food premises, so that EHOs know what food businesses exist or are planned, is now in force.
- In parts of the USA, persons wishing to run a food business are granted a licence after passing a hygiene examination.

See: Food Safety Act 1990.

Lighting

- This should be suitable for the task and the person performing it.
- It is difficult to set absolute figures although 500 lux is sometimes quoted as a good general figure for kitchens, measured at the work surface.
- Detailed work may require up to 2000 lux.
- The requirements of individuals vary considerably.
- Too much light is as bad as too little and the number, choice and placement of light fittings is a specialist task. There is an HSE Guidance Note (HSG 38) on lighting at work.

Advantages of good lighting

Assists cleaning
Discourages pests
Improves performance and safety
Creates a comfortable work environment

Artificial lighting is more controllable than natural light, but the latter may be more pleasant although heat from solar gain is a potential problem.

Lima beans Contain considerable quantities of cyanide-containing substances which must be removed by boiling in an open pan so that poisonous hydrogen cyanide gas may escape. Special low cyanide varieties of these beans have been developed for agriculture and only these are permitted to be grown, for example in the USA.

Lindane Common name for the insecticide gamma BHC; its use is banned in some countries, including Britain.

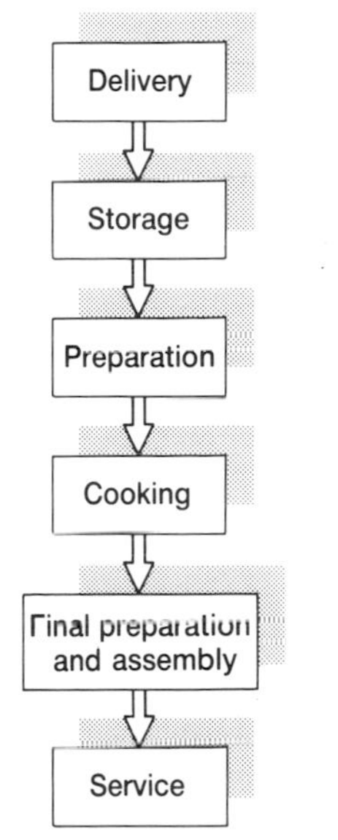

Note: The flows of people, equipment and utensils (including disposable and re-usable containers) and waste must also be considered.

Linear flow (of food)

Linear flow In the design of kitchen and food production units, the flow in one direction from raw materials to finished product.

Advantages

- Cross-contamination is minimised by separating clean and dirty processes
- More efficient cleaning
- Linear operation can be smoother and more economical in time and effort.

A typical line of progress in a kitchen would be
goods inward, storage, preparation, cooking, service, washing-up.

Equipment and work areas need to be carefully arranged so that linear flow is possible. The flow of foodstuffs, persons, equipment and wastes need to be considered. *See*: Design of equipment and food premises.

Lipophilic Literally 'lipid-liking', lipids being fats and oils. A term applied to the tails of detergent molecules.

Liquid egg A product consisting of mixed whole eggs used for convenience in bakeries. It has been implicated in outbreaks of food poisoning and is pasteurised (64.4 °C for 2½ minutes with immediate cooling to 3.3 °C, to pass the alpha amylase test) to reduce the numbers of pathogens such as salmonellae. Being a batch product, one contaminated egg could affect the rest.

Listeria monocytogenes A bacterium, commonly referred to simply as 'listeria', which is widely distributed in nature.

- It has been found in insects, sewage, decaying vegetable matter, soil, ruminants (e.g. cows), humans, all types of farm animals and domestic pets.
- In humans it causes the disease known as listeriosis.
- Young children, the elderly, the sick and pregnant women are most at risk.
- There is a possibility of miscarriage and stillbirth resulting from infection.
- It can cause a variety of problems, including meningitis, septicaemia, and abscesses.

Important considerations

1) The optimum temperature for growth is 37 °C, but it grows at all temperatures down to 2.5 °C. It is capable of multiplying in refrigerators.
2) Foods which have had problems with this organism include commercial chilled foods, such as prepared salads, sandwiches, pates, cooked meats and ready-to-eat meals which just require reheating.
3) Soft white cheeses with a crusty rind, such as Brie and Camembert and unpasteurised milk have also been suspected.
4) Its presence in cooked food is a good indicator of poor hygiene practises during manufacture, distribution or sale.
5) Salad foods should be washed before use.

6) Cook-chill products must not be eaten after the 'use-by' date has expired.
7) Reheating foods must be thorough and to a centre temperature of at least 70 °C.
8) Refrigerators should be kept below 5 °C and defrosted regularly.

The organism is Gram-positive, non-spore forming and has a thermal death time of 10 minutes at 58–59 °C.

Lister, Joseph (1827–1912) Professor of Surgery at the University of Glasgow around 1860 who developed the use of carbolic acid (phenol) as an antiseptic in operations. It resulted in a dramatic reduction in post-operation infections.

Listeriosis The disease caused by the bacterium *Listeria monocytogenes.*

Liver fluke (***Distoma lanceolatum/hepaticum***) Trematodes affecting the bile ducts of pigs, sheep and cattle. They are similar in size and shape to flattened woodlice and their presence causes bile ducts to enlarge, giving the so-called 'pipy' liver. Badly affected livers are unfit.

Load line A marking found on display refrigerator and freezer cabinets indicating maximum recommended capacity. If the level is exceeded the temperature of the food at the top may be above a safe level. *See*: Danger zone.

Local Authority Administrative areas in Great Britain including District, Borough and City councils which deal with various matters including environmental health. Officers including EHOs are appointed by the Local Authority.

Lockers In food businesses, storage for outdoor clothing must be provided. If storage is in a food room, a locker or cupboard is required.

Log phase *See*: Bacterial growth curve.

Lovibond comparator A device using coloured discs, graduated in intensity, which may be compared with the colour of an experimental test. The technique is widely used, for example in milk (Resazurin test) and water analysis (chlorination and pH).

Low-acid food Foods over pH 5, including meats, fish and many vegetables. *See*: Acid foods (classification of), Canning.

Lux A measure of illumination, one lumen per square metre. The lumen is a measure of light received or emitted.

MacConkey agar An enriched agar medium on which coliforms readily grow. It can be used for the isolation of coliforms from water samples. *See*: Presumptive water (coliform) test.

MacConkey broth An enriched and differential broth medium (containing lactose and a pH indicator) used to demonstrate the presence of coliform bacteria in water. *See*: Presumptive water (coliform) test.

Made-up dishes These contain several ingredients and are thus liable to a greater risk of contamination from both ingredients and handling. An example is a meat or fish pie. Proper temperature storage, together with thorough cooking are essential.

MAFF (Ministry of Agriculture, Fisheries and Food) One of two Government Ministries in Britain involved in food matters. *See*: Department of Health.

Maggot Larva of insects such as flies.

Maize weevil (*Sitophilus zea-mais*) Common tropical pests often imported with maize (corn).

Malic acid An organic acid present in apples, and also manufactured. It is used to lower the pH of some canned soups.

Malta fever Another name for the disease brucellosis. *See*: *Brucella abortus*.

Market stalls These premises have separate hygiene regulations which are similar to those for other food premises. *See*: Appendix 1: Food Law.

Margarine A high-fat product similar to butter, made with various oils and fats. It is a low-risk food, but it is perishable, so subject to spoilage. Refrigerated storage is recommended.

Mastitis Inflamed udder in cows, due to a bacterial infection, often affecting milk quality.

Mayonnaise There is a proven risk of salmonella food poisoning, caused by *Salmonella enteritidis phage type 4* from mayonnaise made with raw egg. Commercially produced mayonnaise, made with pasteurised egg, is available.

Meals-on-wheels A system of delivering hot meals by road in vans equipped with heated containers. Temperature control is the main problem since there may be a delay of some hours between loading and delivery.

Mealworm Larva of the mealworm beetle (*Tenebrio molitor*).

Mean generation time The time taken for a population of bacterial cells to go through its life cycle, i.e. to double in numbers. Such times are usually short for prokaryotic cells, compared to eukaryotic cells, and for smaller cells compared to larger cells. (Bacteria are small prokaryotic cells.) For most bacteria the mean generation time (also known as doubling time) is from one to three hours, but some have been shown to double in number in as little as ten minutes. Reproduction is by binary fission.

Measles

1) A term used in the meat trade to describe the cysts (bladderworms/ cysticerci) of the tapeworms affecting pigs and cattle. *See*: Beef tapeworm, Pork tapeworm.
2) A notifiable viral disease of humans.

Measly beef/pork Meat affected with the cysts of tapeworms. *See*: Beef tapeworm, Pork tapeworm.

Meat

- A high-risk food when cooked due to its high protein content.
- Both raw and cooked meats should be be stored in a refrigerator.
- Raw meat or anything which has been in contact with it should not be allowed to come into contact with other food.
- If possible raw meat should be kept in a separate refrigerator to avoid cross-contamination.
- Raw meat or poultry must never be stored above other foods to avoid drip contamination.
- The surface of a piece of raw meat from a healthy animal may be contaminated with micro-organisms introduced during slaughter or butchery.
- The inside of a piece of meat will normally have no harmful bacteria.
- Apart from the risk of tapeworm, rare-cooked meat such as steak is safe as long as the temperature has reached a level sufficient to kill any surface organisms.
- Minced meat is likely to contain bacteria all the way through and must be thoroughly cooked.

See: Beefburgers, Growth requirements (bacterial), Poultry.

Meat products

- Items containing meat such as pies, pasties, sausages, rissoles and many others.

- These are more likely to cause hygiene problems than meat itself since they contain multiple ingredients which are handled considerably during preparation.
- Contamination may be introduced and spread throughout the product.
- They should be stored under refrigeration and thorough cooking is essential. *See*: Centre temperature.

Meat Inspector An authorised officer, with a professional qualification, appointed by a local authority to inspect all food animals before, during and after slaughter, to ensure that they are fit for human consumption. In some countries including parts of Europe, meat inspection is performed by veterinary surgeons.

Meat thermometers A device used for measuring the centre temperature of meat during cooking. The use of such is the only reliable way to ensure that the centre has reached a sufficiently high temperature (minimum 63 °C) to ensure that bacterial cells (other than spores) have been destroyed. Washing and disinfecting of the probe are essential after use, to prevent cross contamination. *See*: Thermometers.

Mechanical dish washing *See*: Two-sink system.

Mediterranean flour moth (*Anagasta kuhniella*) An insect pest of stored cereals affecting milled cereals, nuts and many other foods including dried fruits.

Medical questionnaire A pre-employment document requiring information from prospective food handlers as to any relevant medical condition such as food poisoning or infectious diseases affecting themselves, family or other personal contacts.

Medical screening This may include a medical examination and/or a questionnaire. Faecal samples may be required from food handlers who are suspected of having certain infectious conditions such as food poisoning. *See*: Exclusion of food handlers.

Melons A large and unusual outbreak of food poisoning (caused by the bacterium *Salmonella chester*) occurred in the USA in 1991 due to melons which were contaminated on their surface rind.

Meningitis A serious disease, with inflammation of the membranes (meninges) of the central nervous system (brain and spinal cord). It is one of a number of possible complications for sufferers of listeriosis, particularly in new-born children.

Mercury

- A metal capable of entering the food chain and causing serious effects such as brain damage and foetal deformities in humans.
- Fish, by eating plankton, may take up mercury from industrial waste pollution, this being the cause of the serious outbreak of mercury poisoning in Minamata Bay, Japan (1950s). Cats died after eating contaminated fish. Chickens' eggs were found to contain the metal (through feeding the chickens with scraps which included fish) and human breast milk was also found to be contaminated.

Mesophile A term used to describe those micro-organisms that have a preferred mid-range temperature for growth of 20 to 45 °C. It includes most of the common spoilage and disease-causing organisms.

Mesosome A membranous organelle in some bacteria, formed by infolding of the cell membrane. It is thought to be the site of production of adenosine triphosphate (ATP), which is the useable form of energy in cells.

Metal detectors Devices employed in food factories to detect physical contamination from metal fragments, parts of machines etc. First aid plasters may have an internal metal strip (similar to that used in banknotes) to aid their detection.

Metals Several toxic metals, including aluminium, antimony, cadmium, copper, iron, mercury and zinc have all been known to cause food poisoning. *See*: Individual entries.

Methylene blue test One of a number of indirect methods used for the estimation of bacterial numbers in, e.g. milk, cream and ice-cream. Methylene blue is blue in the presence of oxygen and colourless in its absence. The test shows how quickly any bacteria present in the milk use up the oxygen; the faster the dye loses its colour, the faster the oxygen is used up, and the more bacteria are presumed to be present in the milk. This test was withdrawn in Britain in 1989 for the purposes of milk testing.

Mice

- One of a number of rodent pests which live in or invade buildings.
- The house mouse (*Mus domesticus*) is the most common and is attracted to all foods and water.
- It does not move far from its nesting site to feed.
- A single breeding pair under good conditions may produce 2000 off-spring within a year.

- Mice may carry pathogens on their bodies, and they also excrete them in their urine and droppings.
- They also spoil food with droppings and spillages.
- Other varieties, such as fieldmice, occasionally enter food premises.

See: Pests, Proofing of buildings, Rats.

Microaerophile A micro-organism that prefers a much lower oxygen concentration (2 to 10%) for respiration compared to normal atmospheric concentration (20%). *See*: Campylobacter.

Microbicide *See*: Germicide.

Microbiology The branch of biology which deals with the study of micro-organisms.

Microbistat A general term for an agent which is able to stop or inhibit the growth of micro-organisms, but does not necessarily kill them. Specific agents are called bacteriostats, fungistats, virustats.

Micrococcus

- A group of bacteria, widespread in nature, found in soil, dust and fresh water, and frequently found on the skin of man and other animals.
- They are often found on inadequately cleaned and sanitised food utensils and equipment.
- They are not pathogenic, but they are responsible for spoilage of various foods.
- They are aerobic organisms, some are psychrotrophic (growing well at under 10 °C), and are halophilic (favour an environment with a high salt concentration).
- This latter property means that they can cause spoilage of processed meats, such as ham and bacon, as well as red meat stored at 10 to 15 °C.
- One type (*Micrococcus varians*) is common in milk, dairy products and animal carcases, as well as in dust and soil. It is thermoduric, can survive pasteurisation, and so is able to cause spoilage of milk.
- Some are pigmented, and discolour the surface of foods on which they grow. For example, *Micrococcus luteus* produces yellow discolorations on meat.

Microwave ovens A modern method of cooking which offers the advantage of greater speed in some cases. In terms of food hygiene there is nothing magical about microwaves; they do not kill bacteria directly but are merely a means of heating food. Many problems with microwave ovens are the result of not following the manufacturers' instructions regarding power and heating times.

- Microwaves are high frequency radiations, produced by a magnetron, which causes food molecules to vibrate generating heat.
- Microwaves penetrate up to 40 mm in food and heating may be very fast, although some foods heat up more quickly than others.
- Water absorbs microwave energy faster than ice and this may cause localised cold spots in foods.
- Heat from microwave cooking, if sufficient, will kill micro-organisms but spores and toxins are likely to remain unaffected.
- Commercial microwave equipment is normally of heavier duty than its domestic equivalent.
- A major use of microwaves is in the reheating of pre-cooked food; the standing time allowed for in some recipes is important so that the food is heated to a safe temperature.
- Microwave defrosting can be very quick but the correct setting and time must be strictly followed.

See: Cling film, Cold spots, Defrosting, Runaway heating.

Mildew A type of fungus which is similar to a mould. There are two types:

1) Downy mildews, so called since they form a downy growth on the surface of the host. Many are severe plant pathogens, many are saprophytes or weak parasites. They are able to form spores, require a moist habitat and grow best in rainy or humid weather. They are very common fungal infections frequently found on grapes, beans, melons, peas and potatoes. *See*: *Phytopthora infestans*.
2) Powdery mildews, so called due to the powdery appearance of the surface of the leaves infected. All are obligate plant parasites. They grow on leaves, flowers and fruits of apple, grape, cherry and many other plants. In the early stages of infection, the mycelium looks like a delicate cobweb.

Milk

- Milk and most milk products are protein foods and readily support the growth of pathogens.
- They are considered to be high-risk foods, exceptions being hard cheeses and acid products (pH less than 4.5).
- Raw milk is known to be likely to contain a variety of pathogens, so it

is subjected to a number of forms of heat treatment, including pasteurisation, sterilisation and UHT.

- Some soft cheeses must by law be kept at either less than 8 °C or less than 5 °C.
- Milk and its products must be subjected to strict temperature control, either refrigerated or kept hot as appropriate.
- Brucellosis, food poisoning and tuberculosis among others have all been traced to milk.
- Returnable milk bottles used as containers for other substances have caused problems (creosote, paint and paraffin have been noted) as have bottles containing the eggs of flies.
- Pasteurised milk must pass a plate count and coliform test (less than 30 000 organisms per ml) as well as the phosphatase test. Untreated milk must pass the same tests but have less than 20 000 organisms per ml. The methylene blue test has now been withdrawn.

See: Appendix 1: Food Law, Dehydration, Individual product entries (Butter, Cheese, Yoghurt, etc).

Milk and dairies law There is a considerable amount of law covering production, composition, distribution, heat treatment and labelling of milk and many of its products including other related foods such as eggs. *See*: Appendix 1: Food Law.

Milkstone A hard phosphate-based deposit occuring in the pipework and heat exchangers of milk processing plant. It may harbour bacteria and is removed by an acid cleaning solution. *See*: CIP.

Milky hake A condition of hake, caused by a protozoan (*Chloromyscium thyristes*) which causes a softening of the flesh. Affected fish are aesthetically unfit for human consumption.

Minced meat A meat product with a high risk of contamination throughout, since surface organisms on the original larger pieces of meat are spread throughout the product, including the centre. Items containing minced meat, such as beefburgers, need to be cooked to a centre temperature of over 70 °C to ensure their safety. *See*: Meat, Meat products.

Mist netting Very fine mesh used by licensed pest control operators to trap birds. Protected species are released and others (pests) are killed.

Mites (*Acarina*) Many species are known, some of which affect food. All are very small (around 0.5 mm in length), and tend to be pale-coloured, often appearing pearly-white under the microscope. The flour mite (*Acarina siro*) is most common; it also affects other foods and causes a sour and 'minty' taint. *See*: Cheese mite (*Tyrophagus casei*).

Mollusc Shellfish such as cockle, clam, escallop, oyster, etc having soft bodies usually encased in hard shells. All are high-risk foods and have been known to be responsible for outbreaks of food poisoning. Strict temperature control is essential. *See*: Shellfish.

Monilia A mould responsible for food spoilage. It appears as a pink growth on bread, causes brown rot of apples and it can spoil cereals and flour. *Note*: The organism has also been described under the genus *Neurospora*.

Moths

- Many species of these insects attack food, mainly stored products including flour, nuts and dried fruit.
- Adults and larvae may eat and contaminate food.
- Larvae also spin dense threads of webbing which cause food spoilage and have been known to interfere with the action of milling machinery.
- Large infestations may result in grain heating (from the body heat of the insects) which in turn causes convection currents resulting in moisture movement and hence mould growth in damp regions.
- Control of moths rests initially on good housekeeping, proper storage and stock rotation.
- Fumigation of infested goods is also possible while insecticide-impregnated strips may control flying adults to some degree; badly affected foods are unfit for human consumption.

Motility of bacteria The vast majority of bacteria are unable to move around on their own. They rely on air, water, people, animals, food, etc to carry them from one place to another. Some forms however possess flagella, which enable those which live in a watery environment to move around within their immediate surroundings.

Moulds

- A group of fungi, most of which are aerobic (require oxygen). They consist of a network of fine thread-like filaments called hyphae, which collectively form a massed growth called a mycelium.
- They often appear as white cottony growths, e.g. on fruits, grains and vegetables.
- The coloured spores of mature moulds are responsible for the characteristic colours which become apparent.
- They obtain their food by absorbing solutions of organic material from their environment.
- They reproduce very quickly, by the production and release of spores, particularly under warm, moist conditions.

- Commonly responsible for the spoilage of a wide variety of foods, they are the most common cause of spoilage of bread and most bakery products.
- Some species are known to produce toxins, known as mycotoxins.

See: Aflatoxin.

Mucor A white filamentous mould affecting food, causing spoilage, the spores of which are commonly found in household dust.

Mucous membranes Membranes which line body cavities, such as the entire respiratory system, including the mouth and nasal passages, the urinary, reproductive, and digestive tracts. They secrete a viscous (thick) fluid called mucus, which helps prevent the surfaces from drying out. *See*: *Staphylococcus aureus*.

Mushrooms Large edible fungi, but some species are poisonous to varying degrees (Death Cap and Destroying Angel). If eaten raw they should be washed or peeled, since they are likely to have surface contamination.

Mussels *See*: Shellfish.

Mutualism A kind of symbiosis where both organisms benefit. An example is the bacterium *E. coli* in the gut of humans. It produces vitamin K and some B vitamins, which are absorbed and used by the human body. The bacterium receives shelter and obtains all its necessary nutrients from its immediate surroundings. Many of the resident bacteria in the large intestine demonstrate this kind of relationship. Such organisms may become pathogenic when, for example following injury to the intestine, large numbers of *E. coli* may pass through the peritoneum (a membrane which lines the abdominal cavity), into the peritoneal cavity. Severe peritonitis (inflammation of the peritoneum) may occur, which can be fatal.

Mycelium The entangled cottony mass of hyphae, which forms the basic structure of the fungi. It is visible to the naked eye. In some forms it makes an organised structure, e.g. the stalk of a mushroom.

Mycobacterium bovis A Gram-positive bacterium, a pathogen mainly of cattle and other animals. It causes bovine tuberculosis, which can be transmitted to humans by contaminated milk or food. Pasteurisation and testing of dairy herds for tuberculosis has greatly reduced the incidence of this disease in humans.

Mycobacterium tuberculosis A Gram-positive pathogenic bacterium which causes the infectious disease tuberculosis, mainly in humans. The organism is resistant to drying and chemical agents and it can survive for six to eight months outside the body. It is one of the most common communicable diseases in the world. Transmission is usually via inhalation, but it can be food-borne (for example in milk), transmitted through the gastrointestinal tract or by direct contact with infected material.

Mycology The study of fungi.

Mycotoxins Toxins produced by some fungi. *See*: Aflatoxin.

Myristicin A toxin and hallucinogenic substance found at low levels in nutmeg.

Myxomatosis A disease of rabbits affecting eyes and lungs. It is not known to affect humans, but affected carcases are unfit.

Nailbrush Must be provided for personal washing facilities. Plastic brushes with nylon bristles are recommended so that they can be disinfected. Since they are a possible vehicle of contamination they should be cleaned and disinfected frequently and at the end of each day be stored dry, not in disinfectant solutions. *See*: Handwashing.

Nails The nails of food handlers should be kept short and clean. Nail varnish should not be worn since it may chip off and contaminate food. Biting the nails is potentially hazardous since it may spread staphylococci from the mouth to the fingers, and ultimately to food. *See*: *Staphylococcus aureus*.

Narcotising Substances in a bait, such as alpha chloralose, have been used to render birds unconscious. The method is used to control feral pigeons and starlings, any protected species being released whereas the unwanted birds are killed. It has been used in city areas and around airfields with limited success and mixed public reaction. Alpha chloralose may prove fatal to small animals such as mice, particularly at low temperatures. It achieves its narcotising effect by inducing a type of hypothermia.

National Health Service (Amendment) Act 1986 This piece of legislation removed Crown Immunity from NHS premises in respect of food and safety. Hospital kitchens are now subject to food hygiene law.

National Health Service (Food Premises) Regs 1987 Clarifies the position of health authorities as being both owner and occupier.

Nature, substance or quality Under the Food Safety Act 1990 it is an offence to sell food which prejudices the purchaser by being either inferior, harmful or defective in some other way. The legal term 'prejudice' implies that the person has suffered in some way.

'**Nature**' is applied to food which is different to that requested, such as supplying margarine instead of butter.

'**Substance**' applies if, for example, compositional standards are not met or if foreign bodies are contained in food.

'**Quality**' applies, for example, if food has deteriorated.

Prosecutions are taken by food authorities and sometimes cases have been lost because the wrong term was used in court.

Nausea A feeling of sickness often preceding vomiting.

Nematodes Small worms inhabiting the respiratory systems of food animals. *See*: *Strongylus* species.

Neurotoxic shellfish poisoning (NSP) Caused by the dinoflagellate *Ptychodiscus brevis* which in North America affects clams, oysters and other shellfish. The disease causes numbness in the mouth and general symptoms of food poisoning. The same organism is responsible for the 'red tides' of Florida. *See*: Algae, Paralytic shellfish poisoning.

Neurotoxin A toxin affecting the nervous system, often seriously. Several organisms, including the bacterium *Clostridium botulinum* produce such toxins. They are also a feature of some types of food poisoning from shellfish, including paralytic shellfish poisoning.

Nisin A controlled antibiotic used as a preservative in cheese and canned foods.

Nitrate (Saltpetre)

1) Both potassium and sodium nitrates are used in cured products such as bacon. Both are converted by a reduction reaction to nitrites by enzymes and micro-organisms during the process. Nitrates are also used as preservatives in cheese and pizzas.
2) Used as a fertiliser in agriculture. Excess nitrate may enter the water cycle and cause illness.

See: Nitrite, Nitrosamine.

Nitrite The characteristic pink colour of cured meats is due to a chemical reaction between nitrite and proteins, producing the substance nitrosomyoglobin. Potassium and sodium nitrite act as preservatives and are known to slow the growth of the bacterium *Clostridium botulinum*. *See*: Nitrate, Nitrosamine.

Nitrogen (liquid) *See*: Cryogenic freezing.

Nitrosamines Reactions between protein and nitrite can produce small quantities of these substances, which have been shown by some researchers to be carcinogenic.

Non-ionic detergents Uncharged molecules used in low-foam preparations and some special cleaning agents. *See*: Detergents.

Normal flora A term used to describe a micro-organism that is normally to be found in or on an animal (such as on the skin or in the gastrointestinal tract) without causing disease. (The term 'flora' is a legacy of the original mis-classification of the bacteria as plants.)

Norwalk virus (Norwalk agent) A food-borne, virus-like particle causing gastroenteritis. All age groups are affected. The symptoms typically include nausea, abdominal cramps, diarrhoea and vomiting and last from one to three days. Transmission is by the faecal-oral route. *See*: Food-borne disease, Viruses.

Notices (informal, improvement and emergency prohibition) Following an inspection of a premises by an EHO or other authorised officer, notices may be sent to the owner of a food business.

1) The so-called 'informal notice' is a letter sent indicating any matters which need attention and stating that a re-inspection will take place after a specified period. Such a letter has no legal backing but is used as a low-key method of getting food premises up to standard. It is not used if the situation is serious.
2) An improvement notice is a legal document served to the proprietor of a food business. It states what is wrong, what must be done to put matters right and gives time (not less than 14 days) for any work to be done. There is a right of appeal which must be stated on the notice. If the notice is not complied with then a prosecution is likely.
3) The emergency prohibition notice is issued when a serious health risk exists and may be used to close food premises immediately. It

must be confirmed by application to the court within three days by the court issuing an emergency prohibition order, otherwise compensation may be claimed.

See: Prohibition notice/order.

Notifiable disease In Britain, the occurrence of a number of dangerous diseases must be notified to the Medical Officer of Environmental Health (now known as the Consultant in Communicable Disease Control – CCDC) so that appropriate control measures may be taken to limit their spread.

Examples involving food include: anthrax, cholera, diphtheria, food poisoning, hepatitis, tuberculosis, typhoid fever and undulant fever.

There are many others such as malaria and smallpox which are spread by other means.

Notification of diseases in food handlers Food handlers who are diagnosed as suffering from:

- food poisoning
- typhoid fever
- paratyphoid fever
- dysentery or
- other contagious illnessess,

must be reported to the local authority and may be excluded from food-related work. Food handlers must themselves report any suspected food poisoning to their manager or supervisor who must in turn inform the local authority/EHO.

Nuclear region (zone/body) One of a number of names (also nucleoid, nucleophore) for the bacterial equivalent of the nucleus of a eukaryotic cell. Bacterial cells are prokaryotic and so lack a true nucleus.

Nucleoid/Nucleophore *See*: Nuclear region.

Nutrient agar A nutrient broth solidified by the addition of agar. *See*: Agar-Agar.

Nutrient broth A liquid culture medium used extensively in laboratories for bacteriological work. It is a complex medium made of meat extracts.

Obligate Compulsory/no choice. The term is most often used to describe an organism's gaseous requirements.

- Obligate aerobe: an organism that must have free oxygen present in order to live.
- Obligate anaerobe: an organism that cannot live in the presence of free oxygen.

Obligate intracellular parasite A parasite that can only actively grow and reproduce inside a living host cell of another species. Such organisms can survive for variable periods of time outside the host cell. *See*: Viruses.

Offal Internal organs such as liver and kidney of food animals. These must be treated in the same way as meat since they are high-risk foods.

Offices, Shops and Railway Premises Act 1963 Health and Safety law covering the workplaces stated in its title. Generally the Act deals with working conditions such as cleanliness, overcrowding, heating, ventilation, staff facilities and the guarding of machines.

Oils (cooking)

- These deteriorate during use, especially if overheated, giving rise to off-flavours in food.
- The substance acrolein, having a bitter taste and unpleasant smell, is released by very hot oils.
- Cooking temperatures in the order of 160–180 °C are used, for example in deep friers, and so there is a risk that the surface of large food items will be cooked while the centre temperature remains too low to destroy pathogens.
- Oils are also subject to rancidity.
- All cooking oils should be changed regularly, owing to the production of molecules which are suspected of being carcinogenic.

Onset period The time between consumption of food contaminated with toxins and the appearance of the first signs or symptoms of illness.

Ootheca The name given to the egg case of the cockroach.

Open food Unwrapped or unpackaged food, which should be covered during delivery, storage, save and service to protect it from contamination.

Open wound Any uncovered area of damaged or broken skin such as a spot, pimple, boil, burn, cut, graze or other skin lesion. The covering of such lesions with a a suitable waterproof dressing is compulsory since they provide excellent conditions for the growth of bacteria which may otherwise escape and contaminate food.

Opportunist pathogen A generally harmless organism in its normal habitat. It can become pathogenic in a host who is seriously injured, weakened or ill, or who is under treatment with broad spectrum antibiotics or immunosuppressant drugs. A person's normal flora (bacteria) can become opportunist pathogens under the above conditions.

Optimum pH The pH at which a species of organism grows best. It is not necessarily the maximum pH for growth. As an example, most salmonellae grow best at around pH 7 but will also grow at higher or lower values. *Salmonella* species have been reported in fruit juice at less than pH 4.5. *See*: Growth requirements (bacterial/fungal).

Optimum temperature The temperature at which a species of organism grows best. It is not necessarily the maximum temperature for growth. Most pathogens grow best at 37 °C, but the range of temperatures for growth may be between 5–63 °C. These figures are reflected in the Food Hygiene (Amendment) Regs 1990/91. *Listeria monocytogenes* is known to multiply at even lower temperatures. *See*: Growth requirements (bacterial/fungal).

Organoleptic assessment

- The use of the senses of taste, smell, sight and texture to judge the quality of foods.
- Hearing is occasionally used, as in the characteristically dull sound produced by shaking a bag of unfit shellfish.
- Commercially, this is carried out by trained taste panels who are known to achieve a high degree of reliability.
- Some individuals also have remarkable abilities to detect the presence of chemicals such as pesticides in foods, and have been employed in the food industry in this capacity.

Oriental cockroach (*Blatta orientalis*) The most common species of cockroach in Britain, being originally a tropical insect. It is a dark brown insect which can grow to up to 30 mm in length, with antennae of similar length. Females lay around sixteen eggs in a brownish capsule (ootheca) up to 15 mm long. The eggs do not hatch until external conditions are suitable. *See*: Cockroaches.

Osmophile A micro-organism able to grow in a high concentration of solutes (salt or sugar), i.e. in an environment with a low a_w. *Saccharomyces rouxii* (an osmophilic yeast) has an a_w of 0.62, compared to most (ordinary) yeasts which require an a_w of around 0.9.

Osmosis The diffusion of water from a region of high concentration (i.e. a weak solution) to a region of comparatively low concentration (i.e. a stronger solution), through a differentially permeable membrane. The preservation of foods by the use of salt and sugar utilise this phenomenon. *See*: Food preservation.

Outbreak of food poisoning Two or more cases, a case being any single person suffering from the condition. Outbreaks occur anywhere food is prepared including domestic and commercial premises, institutions and occasional venues such as outside catering. The most common reason for outbreaks in general is lack of temperature control, often made worse by preparing food too far in advance. *See*: Food poisoning.

Outdoor catering The provision of food, often for large numbers of people, in a situation often not purpose-designed, poses many food hygiene problems. Typical events include pop concerts, weddings, barbecues, sports events and fetes.

There is a comprehensive set of guidance notes issued by IEHO which covers all aspects.

- Planning and consultations between organisers and EHOs before the event to identify problems and their solutions.
- Provision of suitable storage facilities including refrigeration.
- Hygienic delivery of foods and equipment.
- Prevention of cross-contamination.
- Temperature control of high-risk foods.
- Personal hygiene facilities.
- Suitable food hygiene training for staff.
- Services such as water supply.
- Refuse disposal.

Temporary facilities should not be allowed to become permanent but should be replaced with proper facilities.

In general the construction and use of equipment and food areas is the same as that for any other food business. An exception is the short-term use of wooden duck boards as a floor surface in tents.

Overlap A measurement of end seams in canning.

Oysters Since these may be eaten raw they are a frequent cause of food poisoning from various organisms including species of *Aeromonas*, *Salmonella*, *Shigella*, algae (DSP, NSP, PSP) and many others. *See*: Shellfish.

Oxygen requirements (of micro-organisms) *See*: Aerobes, Anaerobes, Facultative, Growth requirements (bacterial/fungal).

Ozone A gas which has been reported to double the storage time of loosely packed small, fresh fruits, such as strawberries, raspberries and grapes. It is used at concentrations of 2 to 3 ppm in the atmosphere.

Packaging Materials such as cardboard boxes which may have been contaminated should be removed before foods are placed into storage. Physical contamination from packaging materials including binding and staples is possible, so removal should be away from food. *See*: Freezer burn, Wrapping materials.

Paint Gloss paint is an acceptable surface for certain areas of food rooms including walls, ceilings and doors, etc. If food-quality washable paint is properly applied it is non-absorbent and may be easily cleaned. However, the finish may be short-lived and flaking can become a problem, which may result in physical contamination of food.

Paralytic shellfish poisoning (PSP) Mussels and other bivalves feeding on poisonous organisms (a type of alga, the dinoflagellates, found in plankton) sometimes become contaminated with toxins which affect the human nervous system (hence called neurotoxins). The problem is most likely during warm weather when plankton 'bloom' (as a result of rapid growth) and cause a visible 'red tide'.

Cooking may not destroy the toxins which produce an almost immediate numbing of the mouth and pharynx (throat) with a tingling sensation spreading to the upper body and arms within six hours. This condition is often fatal, death being due to paralysis of the respiratory system.

Three species of alga, *Gonyaulax catanella* (North America and Japan), *G. tamarensis* (USA and Europe) and *Pyrodinium bahamense* (South America and SE Asia) have been identified as the causal agents.

Parasite An organism that lives in or on a living host. It obtains nutrients and shelter from the host and so the host is damaged. Such organisms, therefore, cause harm to the host, resulting in disease.

Parasitism A form of symbiosis in which one species benefits and the other is harmed.

Paratyphoid fever A mild food-borne disease of humans caused by the bacterium *Salmonella paratyphi*. Symptoms are much less severe than those of typhoid fever and resemble those of salmonella food poisoning.

Parev ice Kosher ice-cream, containing no milk fat but at least 10% other fat.

Pasteur, Louis (**1822–1895**) French chemist, professor at Strasbourg and later at Lille, whose work with fermentation showed that spoilage organisms could be destroyed by heat. In 1877 he produced revolutionary evidence proving that the disease anthrax was caused by micro-organisms and so helped lay some of the foundations for modern microbiology and medicine. *See*: Koch, Pasteurisation.

Pasteurisation Technique named after Louis Pasteur which uses heat to destroy micro-organisms in a product, so extending its shelf-life. The process is not severe, for example milk is heated to 72 °C for 15 seconds (80 °C for ice-cream). This is sufficient to kill pathogens but some spoilage organisms survive as do toxins and spores. *See*: HTST, Ice-cream, Liquid egg, Sterilisation, Thermoduric, Thermophilic.

Pastry products Pies, pasties, quiches and similar items which contain high-risk foods such as meat, egg or cheese must be kept refrigerated to limit the growth of pathogens and moulds. Some of these products have a very short shelf-life of only a few days. *See*: Appendix 1: Food Law, Temperature control, Use-by date.

Pathogen A micro-organism capable of causing disease. Pathogenicity is dependent on two main factors:

1) the ability to produce potent toxins
2) the aggressiveness (invasiveness) of the agent.

A successful pathogen need not possess both properties.

Pediococcus A group of Gram-positive bacteria which have a number of important growth characteristics. They:

- are microaerophilic,
- grow in the temperature range 7 to 45 °C,
- are slightly halophilic,
- produce lactic acid as a result of fermentation.

They can cause spoilage of food due to their ability to grow at low temperatures and their ability to ferment sugars can ruin brined vegetables and beers. Some members (such as *Pediococcus cerevisiae*) are used for meat preservation by fermentation, as in products such as smoked sausage and salami. The bacteria are inoculated into the meat, where they release preservative organic acids inside the food.

Penicillins A group of antibiotics produced either by the mould *Penicillium* or they may be made synthetically in the laboratory.

Penicillium Moulds producing characteristics of some cheeses. *P. roqueforti* is used in several blue-veined cheeses while *P. camemberti* coats both Brie and Camembert. These moulds may also cause spoilage of other foods.

Perishable foods All foods deteriorate eventually but this term is normally applied to those with a short life and includes the high-risk foods and some fruits and vegetables. *See*: Appendix 1: Food Law, Best-before date, Storage, Use-by date.

Personal hygiene The means by which food handlers prevent contamination on or within themselves from gaining access to food. People working with food must have a good standard of general health and adopt good habits and standards. There are a large number of legal obligations including wearing clean overclothing, not smoking and reporting food poisoning, as well as an overriding obligation not to cause contamination of food. The following areas all come under the umbrella term of personal hygiene:

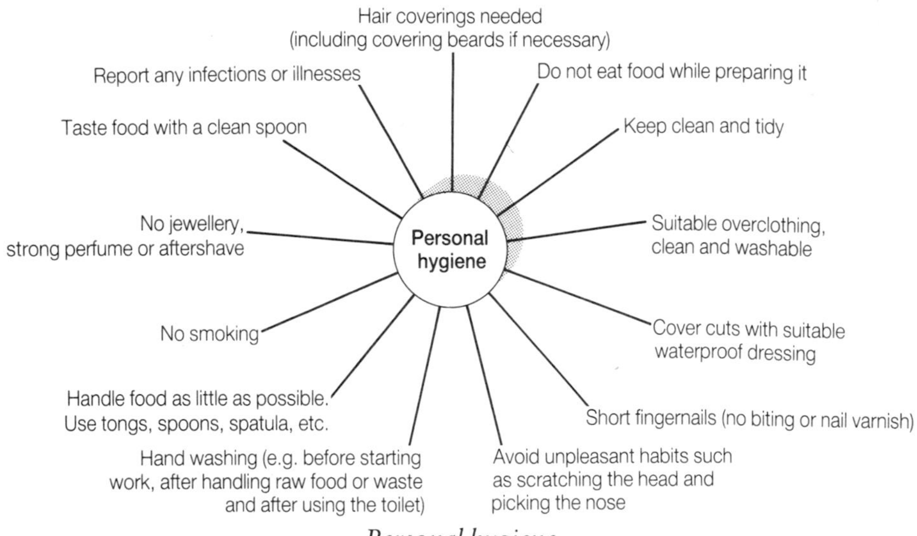

Personal hygiene

Pests and pest control

- The prevention, detection and destruction of insects, animals and birds which attack foodstuffs.

- Common pests include rats, mice and cockroaches, all of which are known to carry pathogens such as *Salmonella*.
- Pests also spoil a great deal of food and cause damage to buildings.
- Generally speaking, the control of pests is a specialist task; DIY is often not successful.
- Control measures are normally physical (prevention of entry of pests, trapping) or chemical (use of pesticides).
- The presence of pest infestations is one of the largest single causes of closure of food premises.

See: Cleaning, Design of food premises, Good housekeeping, individual entries for pests (e.g. Rat), Proofing, Storage.

Pesticides Chemical substances used to kill pests. Poisons used to control rodents (for example warfarin), must be used in food rooms with extreme care by specialists. Pesticide residues on agriculturural produce

Common pests in the food industry

Pest	Signs
Insects	
Flies, beetles wasps, cockroaches, psocids, silverfish, ants	Eggs, larvae, pupae, visual sightings, dead bodies
Rodents	
Mice, rats	Spillage of food around sacks, packets etc; gnawing damage to fabric of building/food packaging materials, smears around pipes, droppings, visual sightings, dead bodies
Birds	
Feral pigeons, sparrows	Droppings, feathers, visual sightings, dead bodies

should be washed off fruits and vegetables before use. A recurrent problem is the eventual resistance of pests to chemical attack and the subsequent search for new control measures. Pesticides may also affect other organisms including humans. *See*: DDT, LC_{50}, LD_{50}, Individual entries for pests (e.g. Cockroach).

Petri dish/plate A flat circular glass or plastic dish, which has a close-fitting lid. Containing various types of agar gels, they are used extensively for the cultivation of micro-organisms in the laboratory.

Pets Animals in general are known to carry a variety of disease-causing organisms which may be passed on to humans directly, or in food, and domestic animals are no exception. There is also a possibility of physical contamination of food with stray hairs and soil. Animals should not be allowed into food rooms. Pet animals may be a problem in domestic premises used as a food business. *See*: Animals, Domestic premises.

pH

- A measure of the acidity or alkalinity of a substance.
- The pH scale runs from 0 to 14 with 7 being the neutral point.
- Acids have a pH of less than 7, acidity increasing as the value gets lower.
- Above pH 7 alkalinity increases.
- Pure water has a pH of 7 and is said to be neutral.

Many foods, including most meats and vegetables have pH values between 5 and 7, such figures being in the range preferred by most bacteria. Below pH 4.5, pathogens do not normally grow. The term pH is derived from the concentration of hydrogen ions (H_) in a substance (power of H). The scale is logarithmic, pH 3 being ten times more acid than pH 4 etc. *See*: Acid foods (classification), Canning, Dark-cutting meat.

Phage An abbreviated form of the term 'bacteriophage'.

Phage typing *See*: Bacteriophage typing.

Pharaoh's ant (*Monomorium pharaonis*)

- A small (2–3 mm) pale yellow ant, originally of tropical origin.
- Inhabits warm buildings such as hospitals, bakeries and kitchens.
- Feeds on a variety of foods: sweet (sugar, jam, honey), protein (cheese, meat) and high fat (lard, butter).
- The small worker ants seem able to discover food sources even in closed containers.
- There is no evidence that this ant carries pathogens internally but their bodies and legs may pick up contamination and transmit it to food.
- Infestations may be persistent since nests are constructed in inaccessible places making treatment difficult.

See: Ants, Juvenile hormone.

Phenol

1) Certain anionic disinfectants, some turning white when added to water, contain phenol as a broad-spectrum bactericide. Due to their strong smell and taste they have no direct food use but are used for toilets, floors and drains. They may be absorbed by the skin producing toxic effects although chlorinated phenols such as hexachlorophene are designed as skin bactericides.
2) During smoking of foods such as fish, phenols and other substances are produced and act as preservatives.

PHLS (Public Health Laboratory Service) Central government laboratory which performs analytical work, gives advice and also publishes statistics relating to, e.g. food poisoning and other diseases.

Phosphatase test Phosphatase is a heat-resistant enzyme found in raw milk. It is an indicator of the effectiveness of heat treatments designed to destroy the bacterium *Bacillus tuberculosis*. The test is chemical and results in a yellow colour if phosphatase is present.

Physical contamination The presence of foreign bodies in food. Many physical contaminants are possible including the following examples:

- pieces of machinery (nuts, bolts, sawblades)
- items from people (pens, buttons, cigarette ends)
- cleaning materials
- packaging (paper, cardboard, staples, string)
- flakes of paint
- rust
- wooden splinters
- insects and rodents (hairs, droppings, bodies).

Bacteria and other causes of food poisoning are not considered to be physical contaminants. *See*: Nature, substance or quality.

Phytophthora infestans A fungus which causes the spoilage of vegetable foods, such as potatoes where it forms a scaly material on the surface, making them inedible. It was this organism which was responsible for the Irish potato famine of 1845.

Pickling

1) A method of food preservation using vinegar which is a weak solution of acetic (ethanoic) acid. Typical products include onions and red cabbage.

2) During bacon curing, the brine used in some processes is called the 'pickle'.
3) A fermentation process used to preserve vegetables employs a weak salt solution which is converted to lactic acid by bacteria such as *Leuconostoc mesenteroides* and *Lactobacillus plantarum*. Sauerkraut and dill pickles are typical products.

See: Food preservation, pH.

Pigeons Wild (feral) pigeons are regular inhabitants and entrants of buildings. They are undesirable in food rooms since they are known to carry pathogens and may cause physical contamination. *See*: Birds.

Pilchard Alternative name for some species of sardine. If canned this food should be treated in the same way as canned salmon. *See*: Fish.

'Pipy' liver *See*: *Fasciola hepatica*.

Plague The bubonic plague caused by the bacterium *Yersinia pestis* carried on the rat flea which swept Europe in the Middle Ages, causing the death of one third of the population.

Plants (poisonous) Food poisoning although most commonly caused by bacteria may be caused by the consumption of many plants. Children in particular sometimes eat poisonous berries. Well known examples of poisonous foods include raw kidney beans, potatoes which have turned green and rhubarb leaves. Non-food poisonous plants include laburnum, holly (berries), deadly nightshade, hemlock and mistletoe. *See*: Alkaloids, Red kidney beans.

Plate freezing A method of freezing flat items of food such as burgers and fish fingers using pressure plates containing a circulating refrigerant. *See*: Frozen food.

Plasmid A small circular piece of DNA in bacterial cells, not located in the nuclear region. It is self-replicating and typically contains a small number of genes. Different plasmids are known, including some which control fertility, some are drug resistant, some have metabolic functions and others (known as virulence plasmids) control the production of some toxins, such as endotoxin and enterotoxin by *Escherichia coli*.

Plasters *See*: Waterproof dressing.

Plastics Have many food uses since they are non-absorbent, non-toxic and may be cleaned effectively. *See*: Chopping boards, Food contact surfaces.

Plesiomonas shigelloides A bacterium which can cause gastro-enteritis; it is associated with the consumption of seafood, including oysters and mussels.

Poliomyelitis ('Polio')

- A disease caused by the polioviruses, which are more stable than most other viruses and can remain infectious for relatively long periods of time in water and food.
- Transmission is mainly by ingestion of water contaminated with faeces containing the virus.
- The virus may cause a sore throat, headache and nausea, but it may affect the central nervous system causing paralysis.
- Control measures include personal cleanliness, adequate heating of food (pasteurisation temperatures are sufficient to inactivate the virus), disinfection of water and prevention of contact with foods by flies.
- Vaccines are available and are highly effective.

Polypropylene A hard plastic material whose major food use is in chopping boards as an alternative to wood. This material is durable and may be easily cleaned although users find that the surface becomes scarred and is somewhat slippery. Sheet polypropylene is also used as a wall covering and in doors (the so-called 'rubber door').

Population (bacterial) growth curve *See*: Bacterial growth curve.

Pork A high-risk food which should be subject to strict temperature control. Pigs are known carriers of *Salmonella* which will be destroyed only by thorough cooking. The risk of infection by tapeworm and roundworm is also avoided by only eating well-done pork although freezing is known to kill both of these parasites.

Pork tapeworm (*Taenia solium*) The risks and precautions for for this are similar to those for the beef tapeworm. This parasite may also form bladder worms in humans which may reinfect the host or infect other people.

Portals of entry and exit The places through which a pathogen enters and leaves the body. For food-borne diseases and food poisoning these are the mouth and anus.

Port Health Authority At ports of entry such as docks, this is the equivalent of the local authority for the purposes of inspection of imported foods and other matters. EHOs are employed by the port authority.

Potable water A term used to indicate that water is fit for human consumption and free from colour, smell, taint and pathogens.

Potassium salts

1) Potassium acetate: a food additive used to preserve colour of plant foods acting as a buffer (stabilises pH).
2) Potassium benzoate: a salt of benzoic acid used as a preservative.
3) Potassium hydroxide (Caustic potash): a component of some oven cleaners used to remove hard, carbonised deposits. It attacks human skin and must be used with care. *See*: Sodium hydoxide.
4) Potassium metabisulphite: a component of Campden tablets, used to stop fermentation in the brewing industry and as a disinfecting agent for equipment used in wine and beer-making, particularly home-made. *See*: Sulphur dioxide.
5) Potassium nitrate and nitrite. *See*: Curing, Nitrate, Nitrite.
6) Potassium propionate: a salt of propionic acid used as a mould inhibitor in bakery and dairy products.
7) Potassium sorbate: a preservative, effective in acid conditions, used to control mould growth (and some pathogenic bacteria) in cheese, bread and bakery products, yoghurt, dessert products, seafood dressing and jams. *See*: Sorbic acid.

Potatoes When exposed to the light or cold a toxic chemical, solanin, is produced in quantities large enough to cause poisoning. Also the green coloured substance chlorophyll occurs and is an indication that the potato should not be eaten. Since most of the solanin is close to the surface, lightly affected potatoes may be deeply peeled and used. Solanin is an alkaloid.

Poultry The main food hygiene problem is that of *Salmonella* contamination of both the flesh and eggs. Intensive methods of production and contaminated feed are thought to be the major problems. The percentage of chickens affected in Great Britain has dropped from around 80% in the 1970s to less than 50% in the early 1990s. Anything which raw poultry or its juices comes into contact with, such as hands, knives, chopping boards, storage containers and other foods is likely to become contaminated. *See*: Cross-contamination, Refrigerators, *Salmonella* (various entries).

Power of Entry An EHO or other authorised officer may enter a food premises at any reasonable time; obstruction is an offence. Samples, documents, photographs and video recordings amongst other things may be taken and unfit food may be seized. *See*: Appendix 1: Food Law.

Prawns *See*: Shellfish.

Premises This term includes all places including ships, aircraft, vehicles, stalls, etc used for food production, preparation or sale as part of a food business. *See*: Food business.

Preparation surfaces *See*: Food-contact surfaces.

Preservation of food *See*: Food preservation.

Preservatives Substances used to stop the growth of spoilage bacteria, moulds and yeasts in foods. A wide range of substances, all controlled by regulations, is used. Some of them are reported to be toxic to humans. Examples of preservatives with food use include acetic (ethanoic) acid, benzoic acid, carbon dioxide, biphenyl, formic acid, fumaric acid, lactic acid, malic acid, nisin, nitrate, nitrite, propionic acid, sorbic acid and sulphur dioxide. *See*: Food preservation.

Preserves Fruit products, similar to jam, with a high sugar content. They are not known to cause food poisoning but are prone to mould growth after exposure to the air, so refrigerated storage after opening is recommended. *See*: Food preservation.

Presumptive water (coliform) test The initial test used to assess water purity. Samples of water are incubated in a specific liquid broth which contains a sugar (lactose). If fermentation occurs, gas is produced. This is presumed to be from the fermentation of the sugar by coliforms. The test does not completely identify the water contaminant as a coliform, since other water-borne bacteria can produce similar results. Any such positive samples then undergo additional tests, called the Confirmed test and the Completed test. If the presumptive test does not result in growth or gas production, the water is assumed to contain no coliforms, was therefore not contaminated with faecal material and is bacteriologically safe to drink.

Prevention of Damage by Pests Act 1949 Legislation which includes:

- a requirement for pest control by the local authority
- notification of infestations by occupiers of buildings
- occupiers to deal with pest infestations
- the control of methods used in dealing with infestations.

There is power of entry under this Act.

Prohibition notice and prohibition order Following a prosecution, if it is considered that there is a health risk, the court imposes a prohibition order that premises, parts of premises, or equipment cannot be used. The prohibition notice stating this is conspicuously fixed to the premises. The order may be lifted after application to the local authority if it can be proved that the risk situation no longer exists.

Prokaryotic A type of cell, possessed by the bacteria, which lack a true nucleus and membrane-bound organelles.

Proofing of buildings In pest control the physical means by which entry into buildings is denied. Examples include close-fitting doors, screens over windows, sealing of all gaps such as those around pipework and maintenance of drainage systems. *See*: Design of food premises, Pest control.

Propionibacter shermanii A bacterium which produces carbon dioxide in large pockets in Swiss cheeses such as Emmenthal.

Propionic acid An anti-fungal preservative agent used in bakery and dairy products. It is often used as calcium, potassium or sodium propionate.

Propyl parahydroxybenzene (Propyl 4-hydroxybenzene, n-Propyl p-hydroxybenzene) A preservative made from benzoic acid with wide use in drinks, pickles, salad cream and fruit pie fillings. *See*: Benzoic acid.

Protective clothing

- Food handlers are required to wear clean and washable overclothing to protect food from contamination.
- Light coloured clothing is preferable since its cleanliness may be easily seen.
- Safety clothing, such as eyeshields when mixing cleaning chemicals, and chain mail gloves if using bandsaws, may also be needed.
- Waiting and bar staff may wear normal clothing.
- The person running the business must supply all such clothing and also provide somewhere to store ordinary clothing away from food.

See: Appendix 1: Food Law, Personal hygiene.

Proteins Proteins are structural materials in living organisms and are an essential part of all living cells including bone, muscle, skin, heart and other organs. Foodstuffs rich in protein are favoured by bacteria and fungi, and so need careful temperature storage. They contain carbon, hydrogen, oxygen, nitrogen, sulphur and phosphorus. The basic unit is the amino acid which is joined with other amino acids by the peptide bond (–OC–NH–) into very large and complex molecules. *See*: High-risk foods.

Protein foods High-risk foods often contain significant amounts of protein upon which harmful and spoilage bacteria feed.

***Proteus* species** A group of actively motile Gram negative coliform bacteria. They are to be found free-living in water, soil and sewage and are normally present in the human bowel and on the skin. Pathogenic species may cause ear, eye and urinary tract infections, as well as 'summer diarrhoea' in children. Some have been involved in the spoilage of meats, seafoods and eggs. *See*: Scombroid food poisoning.

Protozoa A group of single-celled organisms, considered by many to be single-celled animals. They require a moist environment and are found in rivers, stagnant pools and moist soils. They feed on both living and non-living material. Some are capable of producing cysts, which are heat and drought-resistant. The protozoa include the amoebas which live in water and soil, most of which are harmless. Also included is *Entamoeba histolytica*, which causes amoebic dysentery in humans. *See*: *Giardia lamblia.*

Pseudomonas

- A genus of Gram-negative bacteria, common in the environment; they are found in soil, on plants, in water and as part of the normal skin and gut flora of humans as a commensal.
- In humans it can become a pathogen when the resistance of the host is low, or when it becomes involved with a mixture of other infecting bacteria.
- It can infect burns, the urinary tract and can cause abscesses, septicaemia and meningitis.
- As a result of their growth they produce characteristic brightly coloured pigments (deep blue, yellow-green and brown-black) which may diffuse into the surroundings and change its colour.
- *P. aeruginosa* produces both an endotoxin and an exotoxin and the infection leads to the production of blue-green pus while *P. mallei* causes the disease glanders in horses.
- A number of species are responsible for the spoilage of many foods, particularly high protein foods, producing 'off' tastes and odours.
- Effects include slime on the surface of meats and eviscerated poultry, yellow-green fluorescence on fish and green rots of eggs (*P. fluorescens*), and 'off' flavours and colourings in milk.
- All species grow well at refrigeration temperatures and in moist conditions (a_w 0.97 to 0.98).

- They are not readily killed by heat, grow poorly if oxygen is not available and are not particilarly resistant to drying; they are fairly resistant to many of the commonly-used disinfectants.

Psocids *See*: Booklice.

Psychrophile An organism that grows best at temperatures below 20 °C. Many common fungi readily grow at these temperatures, including species of *Cladosporium* and *Penicillium*.

Psychrotroph An organism that grows at temperatures not far above freezing, hence it will grow readily at refrigeration temperatures. Such bacteria are found chiefly in the genera *Alcaligenes*, *Flavobacterium* and *Pseudomonas*.

Ptomaines Substances produced when food decays, thought in the 19th century to be the cause of food poisoning, but later found to be harmless.

Public analyst Laboratory-equipped analyst of food and water appointed by food authorities but also used by others. Many types of analysis including chemical and compositional are possible.

Puffer fish Members of the order *Tetraodontiformes*, such as porcupine fish, ocean sunfish and puffers, have poisonous internal organs; the toxin involved (tetradotoxin) may contaminate the flesh. The disease is often fatal (60% mortality) due to respiratory failure. Other symptoms are numbness in the face and a floating sensation. *See*: Fugu fish.

Pupae (insect) A stage in the life cycle of some insects, between the larva (caterpillar, grub or maggot) and adult. It is the inactive stage, when it is in a cocoon. The adult insect emerges fully-formed from the pupa.

Pus The yellow-white matter produced at the site of an infection, e.g. in boils and pimples. It consists of dead and living white blood cells, dead and living bacteria, bacterial toxins and fluid. If allowed to contaminate food it may lead to an outbreak of food poisoning. *See*: *Staphylococcus aureus*.

Pybuthrin A short-lived low-toxicity insecticide.

Pyrethrum Naturally-occuring insecticide found in the pyrethrum flower (*Chrysanthemum cinerieafolium*) exhibiting very quick knockdown with low toxicity to mammals. The substance is effective for a few days and is broken down by sunlight. Piperonyl butoxide is frequently added to prevent recovery of insects.

Pyrethroids Synthetically prepared insecticides similar to pyrethrum, examples being bioallethrin and bioresmethrin.

Q fever A disease caused by the organism *Coxiella burnettii*, a rickettsia. It produces a pneumonia-type infection, and is rarely fatal. The organism is found in domestic livestock, and animal to animal transmission is by the bites of arthropods. In humans, the disease is most commonly associated with the consumption of unpasteurised milk and through aerosol inhalation in dairy sheds and similar places. Workers in meat and hide processing plants are also at risk. The organism is killed by heat treatment.

Quality assurance (QA)

- A system covering all aspects of production, distribution and end-use of a product.
- Such a system may commence by detailing exact specifications of raw materials and then go on to closely define each stage of a production process.
- At the end of this, if the system has been adhered to, quality of the product should be both satisfactory and consistent.
- Such systems are used to ensure that all legal obligations regarding food safety are met and are also a means to good customer care.
- The EC standard ISO 9000 and the British Standard BS 5750 are frequently used in industry in this context.

See: Due diligence, HACCP.

Quality control (QC) The means of ensuring that product specifications are met. For example, in the food industry bacteriological tests, taste panels and magnetic detectors are frequently used. *See*: Quality assurance.

Quaternary ammonium compounds (QAC or QUATS) Cationic detergent molecules having bactericidal/bacteriostatic qualities. They are both more expensive and limited in their action than hypochlorites but are effective in alkaline solutions and against some moulds.

Quick freezing Produces a better quality product with less drip than slow freezing since water is frozen in place inside cells and is not subsequently lost as drip on thawing. Commercially, quick freezing is most often achieved with small, blast-frozen food items. *See*: Frozen food (production of).

Radiation Particles and electromagnetic waves emitted from various substances and devices, examples being X- and gamma rays used as a means of food preservation. *See*: Irradiation.

Rancidity (hydrolytic and oxidative) The production of substances causing off-flavours and odours in fatty foods such as oils and dairy products.

1) Hydrolytic rancidity involves the release of free fatty acids such as butyric acid which produces the characteristic smell of rancid butter. The reaction is catalysed by lipases (enzymes).
2) Oxidative rancidity is the result of oxygen attaching to double bonds of unsaturated fatty acids forming organic peroxides. Metal such as aluminium foil in contact with foods under refrigeration and freezing speeds up this reaction.

Rats In Britain two main varieties of rat, the black rat (*rattus rattus*) and the brown or common rat (*rattus norvegicus*) are pests of food and other premises.

Main problems

- Rodents feeding on human food are known to carry a number of pathogenic organisms including *Salmonella* and *Leptospira.*
- The bodies are likely to be contaminated due to contact with sewage and refuse.
- Apart from dangers of disease, rats cause damage and wastage to foodstuffs and damage to buildings.
- The incisor teeth of rats must be continually worn down by constant gnawing, the evidence of which is a common first sign of infestation.
- Rat droppings contaminate food.

Control measures

- Rats avoid open spaces and develop 'runs' in covered areas including undergrowth, false ceilings, behind undisturbed food in storage and in cavity walls.
- Prevention of infestations largely rests on good housekeeping to deny food and water.
- Regular inspections to identify early signs.
- Effective waste disposal and stock rotation.
- Proofing of buildings.
- Rodent control is a specialised task best left to experts.

See: Baits, Black rat, British Pest Control Association, Brown rat, Mice, Pests, Proofing, Sticky boards, Trapping, Warfarin.

Raw foods Foods which have not been cooked may contain bacteria which can cause problems due to cross-contamination. Examples include

Salmonella enteritidis from chicken and *Clostridium perfringens* from soil on root vegetables. Raw food should be kept apart from cooked food.

Raw milk Milk which has not been heat-treated by pasteurisation, sterilisation or other means to destroy pathogenic organisms. Cases of food poisoning involving *Campylobacter*, staphylococci and salmonellae have been noted, together with other diseases such as listeriosis, tuberculosis and brucellosis. The law requires appropriate labelling of untreated milk. *See*: Milk, *Streptococcus* species.

Red kidney beans (*Phaseolus vulgaris*) Contain a toxin (haemagglutin or lectin) which affects some people with nausea, vomiting and other symptoms occuring within six hours after eating undercooked beans. Recovery is rapid, normally within one day. The toxin is destroyed by rapid boiling for ten minutes. It may be wise to throw this water away and complete the cooking with fresh water. Cooked canned beans do not require this treatment. Similar toxins may be found in other varieties of beans and pulses which should also be well cooked.

Red whelk (*Neptunca antigua*) Larger, smooth-shelled relative of the edible whelk. It causes occasional outbreaks of food poisoning similar to paralytic shellfish poisoning due to the release of tetramine (a toxin) from its salivary glands.

Refreezing of food Each time food passes through the danger zone of temperatures the numbers of bacteria on it increase. If thawing, reheating and cooling are repeated, the numbers of organisms may build up and become large enough to cause food poisoning. Food must not be frozen more than once.

Refrigerators

- Normal recommended refrigerator temperature is between 1–4 °C; this satisfies the legal requirements for the storage of all high-risk foods.
- High-risk foods must be kept at either less than 8 °C or less than 5 °C.
- In refrigerators most harmful bacteria are dormant and spoilage organisms are slow to multiply.
- The shelf-life of products is extended by some days.

Use of refrigerators

1) Locate away from heat sources.
2) Load correctly to allow a good circulation of air around foods.
3) Hot food should be cooled beforehand.

4) Foods must not be stored in cans except for canned meats; these require only a few hours chilling to facilitate slicing.
5) Containers must be food-grade and covered.
6) It is preferable to have separate refrigerators for raw and cooked foods in commercial premises.
7) If only one refrigerator is available raw meat should be stored below cooked foods to avoid any cross-contamination by drips.
8) Cleaning must be regular, not forgetting the door seals.
9) Temperatures of the product and the air in the refrigerator must be taken and recorded several times per day.
10) Defrosting may be manual or automatic and is essential to avoid build-up of ice which limits efficiency of the appliance.

See: Appendix 1: Food Law, Cook-chill, Growth requirements (bacterial/fungal), *Listeria monocytogenes*, Temperature control.

Refuse (storage and disposal) Accumulations of waste in food areas must be minimised and confined to closed containers, such as bins with close-fitting lids or heavy duty plastic bags on stands. These containers should be regularly (at least daily) removed to a secure place well away from food areas and collected for disposal. Refuse areas, if managed badly, may be a source of food for rodents and animals. They should be maintained and cleaned to a high standard. *See*: Compaction.

Registration The proprietor of an existing or proposed food business, or the owner of the premises in some cases must notify the local authority. Exemptions include those premises used for less than five days in five consecutive weeks and premises and vehicles such as milk and dairies already licensed under other regulations. The object of the regulations is to inform EHOs of the existing and proposed food trade in their area so that they may organise inspection rotas more effectively. There is no charge for registration on the official form, it cannot be refused and need not be renewed. The registration is not a licence. *See*: Appendix 1: Food Law.

Regulations Many Acts of Parliament (such as the Food Safety Act 1990) contain powers for the appropriate Minister to make regulations. These are more detailed requirements than are contained in such Acts. Examples include labelling, training and temperature control regulations. *See*: Appendix 1: Food Law, Individual entries.

Reheating As a general rule, food which has been refrigerated or frozen must only be reheated once, to above 63 °C (70 °C is preferred) or 'piping hot'. This is to avoid the build-up of bacteria if reheating and cooling are repeated. *See*: Refreezing of food.

REHIS (Royal Environmental Health Institution of Scotland) An organisation similar in scope to IEHO.

Reported cases of food poisoning/food-borne disease

- Figures for cases of food-related disease are thought to be unreliable since persons, for various reasons, fail to notify the proper authorities.
- Various estimates from 2 to 10 million cases per year contrast sharply with the official figures of under 60 000 reported in the early 1990s in Britain.
- The reporting system itself may be faulty and people with short-lived symptoms may not go to a doctor; mis-diagnosis is also possible.
- Food handlers must report such suspected illnesses. *See*: Exclusion of food handlers.

Reporting of Injuries, Diseases and Dangerous Occurrences Regs 1985 These regulations require employers to report all serious accidents including those which result in employees missing more than three days' work, acute illnesses and some other matters including explosions and collapse of structures.

Resazurin A colorimetric test carried out on raw milk which determines the bacterial load. Resazurin is a violet dye which turns pink and eventually becomes colourless. The time taken for this colour change indicates the number of organisms present.

Reservoir of infection The common location where a pathogen may always be found in considerable numbers. Examples include humans as

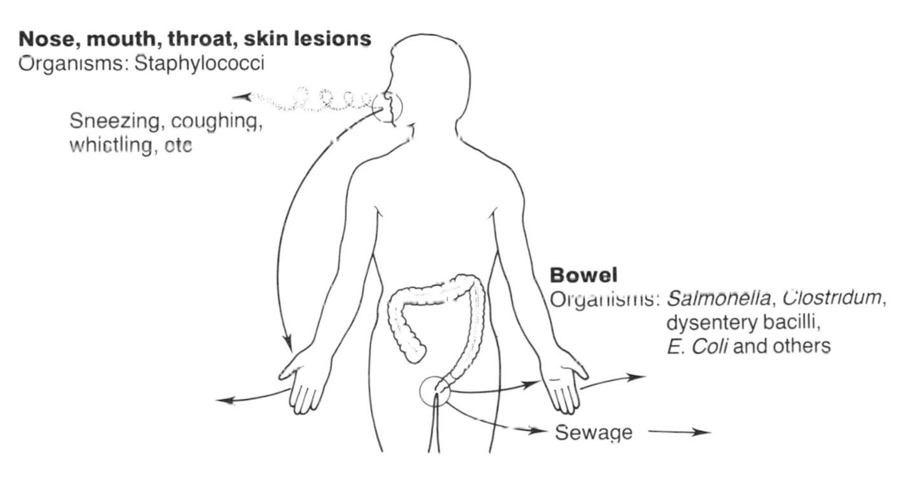

Human reservoirs of infection

reservoirs of *Staphylococcus aureus* and poultry as reservoirs of *Salmonella*.

Resident organisms Those commonly living permanently in a particular location, for example staphylococci on the human skin.

Residual insecticide Substances toxic to insects over a long-period, examples being vapour-emitting strips and lacquer-based insecticides. *See*: DDT, Insecticides.

Rhizopus A mould causing food spoilage, particularly in humid and damp conditions affecting berries, fruits, vegetables and bread.

Rhubarb Leaves contain toxic substances (such as oxalic acid) and should not be eaten.

RIDDOR *See*: Reporting of Injuries, Diseases etc Regs 1985.

Rice

- Cooked rice, due to the risk of *Bacillus cereus* food poisoning, should be kept under strict temperature control; this means either refrigeration or keeping it hot.
- It is included in the list of high-risk foods.
- Typical risk situations include take-aways and rice salads in buffets.
- The practice of adding a fresh batch of cooked rice to the remains of the previous one will allow any contamination present to enter the fresh batch.

Rice weevil (*Calandra* or *Sitophylus oryzae*) A pest of stored cereal products including wheat and rice. Adults have a characteristic 'snout' and 'elbowed' antennae and are chestnut-coloured insects up to 4 mm long.

Richmond Committee An advisory committee reporting on the microbiological safety of food to government ministries involved with food. The reports are very comprehensive and make detailed reference to a wide variety of issues including food-borne illness, HACCP, legislation, manufacturing processes and poultry meat. The reports make many conclusions and recommendations some of which have been taken up and some, including compulsory licensing, ignored as yet.

Rickettsia A micro-organism which is similar in some respects to both bacteria and viruses, causing diseases such as typhus, Rocky Mountain spotted fever and Q fever. All of these are spread by lice, fleas and ticks but there is evidence to implicate raw milk from sheep, goats and cows in their transmission. *See*: Raw milk.

Rinse (hot, after washing up) Water at 82–88 °C is an effective final rinse for utensils and equipment since it will disinfect, remove any traces of chemicals such as detergents and make the clean item hot enough to air dry. This avoids the problem of contamination from cloths. If a high enough temperature cannot be reached then a sanitiser may be added.

Rodenticide A poison capable of killing rodent pests such as rats and mice.

Rodents

Rodents Animals including rats, mice, squirrels and beavers. Rats and mice are significant pests and both are carriers of disease-causing organisms. Their exclusion from food premises is a high priority.

Roe The mass of eggs in a female fish (hard roe) or the milt (reproductive gland) of a male fish (soft roe). They are both high-protein foods and thus high-risk foods.

Rolled (stuffed) joints These have caused problems since bacteria previously on the outside of the meat, together with those introduced from the food handler or other food items, are transferred to the centre during preparation. They may survive cooking with *Clostridium perfringens* (a spore-forming bacterium) being a particular problem. *See*: Centre temperature.

Roller drier A device used for dehydrating 'paste' products including mashed potato or liquids such as milk. It consists of a slowly rotating, steam-heated drum upon which the product is spread, being removed when dry by a scraper blade. *See*: Dehydration, Food preservation.

'Rope'

1) In bread, a condition caused by the heat-resistant bacterium *Bacillus subtilis* producing a sickly odour and soft, cloying texture.
2) In milk, a slimy condition caused by the bacterium *Enterobacter aerogenes* (*Bacillus aerogenes*).

Rotaviruses The most common cause of viral food poisoning in Britain. *See*: Viruses.

Roundworms (parasitic) Many of these affect food animals and fish, examples being:

1) *Ascaris lumbricoides* causes 'milk spot' in the livers of pigs and sometimes has severe effects on the growth of these animals.
2) *Oesophagostomum radiatum* causes decomposition of the intestines of cattle and sheep.
3) *Oncocerca gibsoni* nests in beef flesh.
4) *Trichinella spiralis* cysts in the muscle of pigs will develop in humans after consumption of undercooked pork.
5) *Anisakis simplex* affects fish and humans.

Routes of contamination or infection The path along which bacteria or other organisms travel from one area or food to another. As an example:

1) *Salmonella* may be present in contaminated feed,
2) enter a chicken,
3) be spread by a wiping cloth to a slicing machine,
4) contaminate cooked ham,
5) multiply in warm conditions,
6) cause food poisoning after consumption.

See: Cross-contamination, Vehicle.

RIPHH (Royal Institute of Public Health and Hygiene) A professional body based in London, England which offers various food hygiene training courses and has an interest in other food and health matters.

RSH (Royal Society of Health) A professional body based in London, which offers various food hygiene training courses and has an interest in other food and health matters.

Rubber gloves Those used in connection with food should be brightly coloured for high visibility, in case of physical contamination. Manual washing up requires gloves if the final rinse temperature is at 82–88 °C. Disposable gloves may be used to prevent hand-contact with foods during production, service or sale. All gloves may become vehicles for bacteria and must be discarded if their condition becomes unacceptable.

Runaway heating A problem in microwave cooking or reheating in which localised 'hot spots' occur. These can be minimised by turning the food during cooking, allowing standing time after cooking and heating only one type of food at a time, since different foods cook in different times. Cold spots are a related problem.

Rust Iron oxide caused by the action of water on iron or steel. A rusty (corroded) surface, of a machine or utensil for example, cannot be cleaned properly and may cause physical contamination of food. A rusted can should be discarded since pin holes may have been made through which the contents may have become contaminated.

Rust-red flour beetle (*Tribolium castaneum*) A common pest of stored cereal products. *See*: Beetles.

Rust-red grain beetle (*Cryptolestes ferrugineus*) A minute insect (3 mm in length) thriving in damp stored foodstuffs. *See*: Beetles.

Safety *See*: Health and safety at work etc (Act) 1974.

Salads Prepared salads are considered to be high-risk foods and should be refrigerated (1–4 °C) or eaten within four hours of preparation. Salads containing fruit, rice or coleslaw must be treated in the same way. The legal minimum temperature is 8 °C (5 °C after 1 April 1993 for some items including rice). *See*: Appendix 1: Food Law.

Salad bars There is a danger of product contamination both from public access and from the practice of 'topping up'. The same controls on temperature apply as for salads.

Salmon If fresh, it must be subjected to strict temperature control, as it is a high-risk food. If canned (as with other canned fish and meat) it must be treated as a high-risk food once the can is opened, and the contents must be stored in a refrigerator, in a suitable covered container. *See*: Fish.

Salmonella

General information

- A very large group of bacteria, consisting of over 2000 types (serotypes), all of which are able to cause illness in humans or animals.
- A case of food poisoning caused by a strain of *Salmonella* is known as salmonellosis.
- In Britain most outbreaks are caused by *S. typhimurium* or *S. enteritidis*.
- The *Salmonella* group as a whole have been consistently responsible for the majority of cases of bacterial food poisoning over many years.
- Salmonellae are non-spore formers, so a temperature of 70 °C is usually sufficient to kill them.
- They are Gram-negative, rod-shaped, facultative anaerobes growing between 5 to 45 °C and cause food poisoning by infection.

Food poisoning by salmonellae

- Large numbers (possibly millions) of bacteria are needed to cause illness.
- Usually the organism has grown in the food to produce the high numbers needed.
- The onset period for the disease is usually 12 to 36 hours, but it may, unusually, be 6 to 72 hours or longer.
- Symptoms are fever, headache and general aching of the limbs, as well as diarrhoea and vomiting. They are normally so severe that the disease is often reported to a doctor.
- The duration of illness is 1 to 7 days.
- Fatalities generally are uncommon, but the death rate is high among infants and the very old.

Foods commonly involved

- Meat and poultry and their products
- Raw milk and milk products
- Eggs.

Contamination of food may be directly or indirectly from:

- human or animal excreta
- all farm and wild animals
- birds
- rodents

- all domestic pets
- terrapins
- cockroaches.

Control of *Salmonella*

1) Prevention of contamination during food production by ensuring that animal feedstuffs are safe.
2) Prevention of cross-contamination from raw foods such as chickens to other foods during delivery, storage and preparation.
3) Disinfection of all items which have come into contact with raw food likely to contain *Salmonella*.
4) Thorough cooking of foods such as eggs, poultry and meat.
5) Effective pest control and restriction of contact between animals and food or food handlers. *See*: Cross-contamination.

Salmonella ealing The organism responsible for an outbreak of food poisoning in infants in Britain (1985). The food involved was dried milk for babies. The company producing it went out of business shortly after the incident.

Salmonella enteritidis phage type 4 (PT4) An organism which has become of major importance in Britain due to the large number of outbreaks of food poisoning in the late 1980s/early 1990s from *Salmonella*-contaminated eggs. *See*: Eggs, shell (hen).

Salmonella napoli Found to be responsible for a large outbreak of food poisoning in Europe in 1982, resulting from eating contaminated chocolate. This was an unusual event in that confectionery products do not normally support the growth of bacteria because of their high sugar content.

Salmonella paratyphi Causes the disease of humans, called paratyphoid (or enteric) fever. It is less severe than typhoid fever and is rarely fatal. Symptoms begin with chills and resemble those of normal salmonellosis. Transmission is similar to that of *Salmonella typhi*.

Salmonella typhi

- The bacterium which causes the serious, often fatal human, food-borne disease called typhoid (or enteric) fever.
- The onset period for the disease is from 1 to 3 weeks.
- Symptoms include malaise (a general feeling of body discomfort), an enlarged spleen, slow pulse and 'rose spots' on the body.

- The intestinal walls may become ulcerated and the organisms become distributed all over the body and if they localise in the gall bladder or kidney, they may be excreted in the faeces or urine by carriers over many months or years.
- Persons who are tested positive for typhoid fever should have a lifetime ban on working with food.
- Foods often involved are milk and egg products, but the main spread is by water contaminated with sewage.
- Efficient sanitary control by sewage removal and treatment and water purification are the best control measures.
- Outbreaks often occur following natural disasters such as floods and earthquakes.
- *S. typhi* is Gram-negative and non-sporing.

Salmonella typhimurium Responsible for many outbreaks of salmonellosis. It is commonly found in the intestine of many animals and birds, especially poultry. The excreta of these animals is therefore potentially able to cause food poisoning, if allowed to contaminate food.

Salmonellosis A case of food poisoning caused by an organism of the *Salmonella* group of bacteria.

Salting A method of food preservation which causes death of bacteria by osmosis. Meat, fish and vegetables may be so preserved. The technique works since water in bacterial cells is removed due to the high concentration of salt ouside the cells. In osmosis, water moves from the higher concentration of water (bacterial cell cytoplasm) to the lower concentration of water (salt). *See*: Curing, Food preservation.

Sampling of food Environmental health departments and others (such as Trading Standards Officers) regularly check the quality and composition of foodstuffs by a sampling programme involving laboratory analysis. Samples may be purchased informally to detect any problems, and then formally if results indicate the need for this.

The way in which formal sampling is carried out involves a legally set procedure since results from analysis may result in prosecutions and appeals. If a formal sample is taken the person from whom it is taken must be given part of the sample so that an independent analysis may be performed if desired as a cross-check. *See*: Bacteriological standard.

Sanitary conveniences

- All premises where persons are employed, including food premises, must have suitable and sufficient sanitary accomodation.

- This must be constructed so that it may be kept clean and be well-lit and ventilated.
- WC compartments must not communicate directly with food rooms or other rooms in the workplace and must not be used for any other purpose such as storage.

As a rough guide, one WC each for males and females per 15 persons is needed. Exact numbers are contained in the regulations covering this matter. *See*: Appendix: 1 Food Law, Personal hygiene.

Sanitiser A specially formulated cleaning chemical capable of both detergent and disinfecting actions. *See*: Cleaning, Disinfectant.

Saprophyte/Saprotroph An organism that lives off dead and decaying organic matter. Such organisms secrete enzymes externally into their surroundings, which digests their food, and the digested substances are reabsorbed back into the cell. Many bacteria and fungi are saprophytes, and so are able to spoil food.

Sarcotases Blood-engorged parasites occasionally affecting fish.

Sardine Also called the pilchard, a member of the herring family. As with other canned fish and meat it must be treated as a high-risk food once the can is opened and the contents should be stored in a refrigerator in a suitable covered container.

Sauerkraut A fermented, salted, pickled cabbage utilising the ability of the bacteria *Leuconostoc mesenteroides* and *L. plantarum* to produce lactic acid, which acts as a preservative. Both of these are good examples of non-pathogenic organisms used in food production.

Sausage A meat product containing cereal binders and spices contained in skins, made from animal intestines or synthetic materials. Some sausages are cured, smoked or cooked. All are generally thought of as high-risk and should be subject to strict temperature control. This means refrigerated (1–4 °C) or hot (over 63 °C). *See*: Appendix 1: Food Law, Meat products.

Saveloy A type of sausage subject to the same risks and treatment as other similar items.

Saw-toothed grain beetle (*Oryzaephilus surinamensis*) Similar to the merchant grain beetle, a pest of stored products, principally cereals. *See*: Beetles.

Scampi (*Nephrops norvegicus*) Also called the Dublin Bay prawn, Norway lobster or langoustine. *See*: Shellfish.

Scavenger An animal which feeds on dead and decaying animal or vegetable matter. It is likely to be exposed to, and therefore may carry, a large and varied population of micro-organisms, including pathogens. Common examples of scavengers associated with contamination of food include rodents and insects (particularly flies). *See*: Eggs, shell (ducks).

Schistosoma The blood flukes such as *Schistosoma mansoni* and *S. japonicum* are transmitted to humans and other animals by infected water supplies. The resulting disease is called schistosomiasis.

Scombroid food poisoning

- A food-borne illness resulting from the ingestion of spoiled fresh fish belonging to the scombroid group, such as tuna, mackerel and skipjack.
- The onset period for illness is from 5 minutes to 2 hours and lasts for usually less than 6 hours. Weakness and fatigue may persist for 24 hours.
- The symptoms resemble those of botulism: headache, dizziness, nausea, sometimes diarrhoea and vomiting, swelling of the face, burning of the throat, difficulty in swallowing and throbbing of the carotid (neck) and temporal (temples) blood vessels.
- Fish contaminated with the toxin (scombrotoxin) have a characteristic bitter or peppery taste.
- The symptoms are caused by the release of a histamine-like substance into the fish, possibly as a result of the activity of a species of *Proteus*, a spoilage bacterium.
- Prompt refrigeration of the fresh fish prevents the occurrence of this illness.
- It is likely that this disease is often misdiagnosed as a food allergy.

Scombrotoxin *See*: Scombroid food poisoning.

Scoops (ice-cream) *See*: Ice-cream.

Scouring powder An abrasive cleaning agent often containing detergent. The use of abrasives may cause minute surface scratches and thus provide harbourage for bacteria.

Screening

1) Medical screening of food handlers involves taking faecal samples and analysing these for the presence of disease-causing organisms. *See*: Exclusion of food handlers.

2) Screening of entrances into food premises with mesh or netting is a means of proofing buildings against pests such as birds.

Seafood A general term often used to describe shellfish.

Seasonal variations of food poisoning

- In Britain there has been a general annual increase in food poisoning outbreaks during the Christmas and New Year period and at Easter, due to increased purchases of poultry (often frozen).
- Similar increases have been noted during the summer months due to warmer temperatures. Strict temperature control is vital during these months.
- The use of casual, often untrained staff at these busy times has no doubt contributed to these increases.

'Sell-by' date A date-coding which shows the last possible date for sale of a foodstuff. The date is of limited value since it does not indicate the shelf-life of the product to the consumer who may assume that there are some days' 'grace' in which to use the food. The term should not now be used. A more useful marking is the 'use-by' date which states a definite end to the recommended safe life of the product.

Sepsis The growth of pathogenic bacteria in living tissue.

Septic A wound which has become infected with living pathogenic bacteria. Characteristically it results in swelling, redness, pain and the production of pus. *See*: *Staphylococcus aureus*.

Septicaemia A condition characterised by actively reproducing pathogenic micro-organisms, together with their toxins, in the blood. It is commonly referred to as 'blood poisoning'.

Serial dilution A laboratory method of determining the actual numbers of bacteria present in or on a substance, such as food. It involves preparing a series of dilutions, each time by a factor of ten. Chosen dilutions are plated out onto nutrient agar dishes, which are then incubated, and the colonies counted.

Serological typing

- A test carried out in order to identify a specific type or strain of an organism, for example a *Salmonella* bacterium.
- Routine staining and physiological tests fail to identify them since they are very closely related.

- Serum is obtained from an animal previously inoculated with a known strain of the bacterium; the serum will contain antibodies specifically against that strain.
- If the serum is mixed with *Salmonella* isolated from a patient suffering from suspected salmonellosis, then the resulting antibody-antigen response makes it possible to identify the strain of *Salmonella* present.
- It may take some time before a positive identification can be made, since there are over 2000 different serotypes of *Salmonella* known.

Sequestering agents Chemical susbstances which render ineffective other sustances by combining or complexing with them, an example being the chemical removal of the substances responsible for water hardness.

Services Mains water, gas and electricity supplies, sewage and refuse disposal as supplied to premises.

Sewage

- The watery material carrying refuse and wastes in a drainage system or in sewers.
- It consists of rainwater, grit, leaves, industrial effluents (such as metals, solvents, oil, grease and food waste), household wastes (including disinfectants and detergents), food scraps, urine, faeces and toilet paper.
- Over 99% of sewage is water.
- Raw (untreated) sewage often has a high concentration of harmful bacteria (pathogens) of faecal origin.
- Rats and cockroaches using sewers, or flies from sewage works may easily pick up contaminants on their feet and bodies and spread this to food.

Sewage contamination The main health risks associated with sewage contamination of foods are as follows.

1) From a range of bacteria, viruses and protozoa that have survived the sewage treatment processes, or from foods prepared using untreated water.
2) There is also the risk of food poisoning through eating raw shellfish, such as oysters, harvested from sewage-contaminated water.
3) If untreated domestic sewage is used to fertilise crops there is a chance that the raw plant foods will be contaminated with various harmful bacteria.

Shelf-life The period for which a product lasts in good condition during storage. This may be from a few days (pasteurised milk) to many years (canned meats). Various dating systems are in use. *See*: 'Best-before' date, 'Sell-by' date, 'Use-by' date.

Shellfish A term used to cover a wide range of creatures (molluscs) including:
- oysters, cockles and mussels, etc (bivalves)
- winkles and whelks (univalves)
- prawns, lobsters, scampi and shrimps, etc (crustaceans).

In the UK around 2% of all foodborne illnesses involve seafood but this rises to over 70% in Japan.

Problems
- Some shellfish thrive in sewage-contaminated waters, for example in river estuaries and coastal waters, and may take in any bacteria and viruses that this water contains.
- The bivalves, being filter feeders, are most likely to do this, so they must be purified before human consumption by immersion in disinfected or clean water for some time.
- Toxic chemicals from industrial wastes may also be absorbed by shellfish, but these (and toxins and viruses) may not be removed by purification processes (see below).
- There is an increasing risk of viral food poisoning from these foods.
- Some people are allergic to shellfish.

Use of shellfish
- The smell of shellfish is a good guide as to their freshness since they release obnoxious chemicals when decomposing.
- All raw shellfish must be kept away from other foods to avoid cross-contamination.
- They should be refrigerated whether raw or cooked at 1–4 °C.
- Rapid and thorough cooking is essential.
- Shellfish must be purchased through a reputable supplier since there is a real danger of contaminated foods from unreliable sources, including those gathered direct from the sea or river estuaries.

See: Algae, Hepatitis A, Metals, Paralytic shellfish poisoning, *Salmonella*, *Vibrio parahaemolyticus*.

Shellfish (purification)
1) Depuration: a technique where shellfish are allowed to self-cleanse by immersion in disinfected water for 36–48 hours.
2) Relaying: shellfish are put into non-polluted water for 4–6 weeks before harvesting.

Both methods are capable of worthwhile reductions in bacteria but are not effective in lowering levels of viruses, heavy metals, toxins and pesticides.

Shelving In common with all other fittings in food rooms, shelving should be constructed from an impervious material which can be easily cleaned. Painted or varnished wood, plastics and stainless steel are suitable if kept in good condition. *See*: Design of food premises.

Shigella

- A group of bacteria found in the intestine of humans and other animals, responsible for the disease known as bacillary dysentery or shigellosis.
- It is more common in tropical countries and has many local names including 'Delhi belly' or 'Gippy tummy'; some cases of traveller's diarrhoea may be cases of bacillary dysentery.
- There are four pathogenic species: *S. sonnei* (which causes a mild dysentery), *S. dysenteriae* (which causes severe dysentery and prostration) *S. flexneri*, and *S. boydii*; all are facultative anaerobic rod-shaped organisms.
- The incubation period is from 1 to 2 days, symptoms including diarrhoea and vomiting, abdominal pain and cramps and, in severe cases, blood and pus in the faeces. Infections are usually limited to the large intestine.
- There is some toxin production (*Shigella dysenteriae* produces toxins, but invasiveness is the more important factor.

Control measures

- The disease can be water-borne, so efficient sewage and water treatments are necessary.
- It is often spread from person to person by the faecal-oral route, particularly amongst children. Good personal hygiene helps prevent this.
- Contaminated food, other objects and flies may also spread the disease; control of flies, and careful preparation of food all help to prevent outbreaks of the disease.
- The most likely route is person-to-person contact in crowded populations that have poor personal hygiene and inadequate water and sewage systems.

Shigellosis A general name for bacillary dysentery, caused by species of *Shigella* bacteria.

Ships These, when used for the carriage of food, are subject to the same general design, construction and use criteria as other food premises.

Similarly, food preparation and service on board ship must follow normal hygienic principles.

Shrimps *See*: Shellfish.

Silverfish (*Lepisma saccharina*) Silvery coloured insects around 10 mm in length having a characteristic pointed appearance with two long antennae and bristly 'tails'. They are of nuisance value and tend to infest badly-maintained crumbling surfaces, such as decaying brickwork, feeding on food scraps and paper. They are nocturnal and thrive in damp conditions. *See*: Firebrats.

Sinks

- In food premises separate sinks must be provided for hand-washing.
- These must not be used for any other purpose such as food preparation or washing up.
- Since hands should be washed after using toilets, the separate sink is a means of limiting the spread of any bacteria which may have contaminated the hands from the intestines.
- The sink must be constructed so that it can be kept clean and maintained in this condition.
- Sinks for handwashing must have hot water, soap, nailbrush and suitable drying facilities. *See*: Hand-washing, Personal hygiene.

Skimmed milk Fat-free milk.

Skin

- Unbroken skin presents an impervious barrier to micro-organisms.
- It is thick, dry and scaly, and its secretions are weakly bactericidal.
- Healthy skin however carries large numbers of bacteria, both permanent (resident) and temporary (transient) species.

Resident organisms, which live on the skin surface, and in cracks, hair follicles and pores, include the bacterium *Staphylococcus aureus*. These are almost impossible to remove by hand washing.

Transient organisms are picked up through contact with the surroundings and from the intestines through poor personal hygiene after using the lavatory. They are fairly easy to remove from the skin by proper hand washing.

Any infections cause extra hazards since bacteria will grow and multiply extensively in the warm, moist conditions found in lesions such as cuts.

Slaughterhouses Premises licensed for the slaughter of food animals. There are comprehensive regulations for the design, construction and operation of these.

Slicing machines The main food hygiene problem concerns cleanliness, particularly between uses for different foods, and the risks of cross-contamination. These machines are also legally defined as being dangerous and cannot be cleaned by young persons (under 18 years) or by persons who have not been trained.

Slime layer An alternative name for the bacterial capsule, the outermost layer of some bacterial cells. It helps to confer virulence on pathogenic forms.

Smears Greasy marks left by rats around pipework, holes or along walls with which their bodies have been in contact. To an expert, the presence of smears is an indication of rodent movement and may, for example, give useful information regarding trap location or proofing.

Smoke point The temperature at which oils and fats start to chemically break down releasing acrid-smelling smoke. The oil or fat becomes contaminated with unpleasant tasting and smelling substances such as acrolein which can affect the taste and smell of food. Smoke point is in the region of 200 °C.

Smoked foods

- Traditionally prepared by hanging foods such as fish, bacon, sausages and cheese over smouldering hardwood wood chips.
- Smoked meats and fish are high-risk foods and should be referigerated (1–4 °C) during storage.
- Foods may be salted or partially cured before smoking. The process produces flavour changes with substances such as phenols in wood smoke having a bactericidal, and thus preservative effect.
- The heat of the process also causes some drying and most non-spore forming bacteria are killed, but *Clostridium botulinum* and others may survive.
- The process is time-consuming (taking several days) and a modern development has been in the addition of smoke-flavouring to manufactured foods; such additions have little effect on micro-organisms.
- There is some evidence to suggest that smoking produces carcinogenic chemicals.

Smoking (by food handlers) Smoking is banned in food rooms.

- The main reason is the contact between mouth and fingers, and then fingers to food; any bacteria in the mouth may be transferred to food in this way.
- Smoking also causes coughing which may spread bacteria from the mouth and nose over a wide area.

- There is a possibility of physical contamination of food with ash, cigarette ends and matches.
- Smoking is increasingly thought of as being anti-social and may create interpersonal problems between staff or others. This is not conducive to good hygiene since quality of work is known to be better in a pleasant environment.
- There are other risks, including risks of fire.

See: Personal hygiene, *Staphylococcus aureus*.

Sneeze screens Barriers between open food and people, designed to protect against the aerosols (probably laden with bacteria) created by sneezing. *See*: Personal hygiene.

Snood A covering for long hair/beards designed to prevent stray hairs from entering foods.

Snuff (taking of) This is not allowed in food rooms for reasons similar to those for the banning of smoking.

Soap Bars of soap have been shown to support the growth of bacteria and may thus pass on contamination from person to person. The preferred soap for food handlers to use is an unperfumed liquid soap in a dosing container. Germicidal soaps are also available. Soap is a cleaning agent made by treatment of fats and oils (e.g. palm oil and olive oil) with heat and alkali (e.g. caustic soda) producing chemicals such as sodium stearate and palmitate. *See*: Detergents, Soapless detergents.

Soapless detergents Detergents such as sodium alkyl benzyl sulphonate used in various cleaning chemicals such as washing up liquid. They do not form scum in hard water and are of neutral pH. Their bactericidal effect is limited unless used in hot water.

Sodium benzoate A preservative with wide use in manufactured foods. *See*: Benzoic acid.

Sodium carbonate Washing soda, a mildly alkaline cleaning agent.

Sodium hexametaphosphate (Calgon™) A chemical used to remove the complex ions that cause water hardness, to produce soft water.

Sodium hydroxide (caustic soda) A strongly alkaline substance capable of dissolving many materials, including hair and burnt-on food. It is a common constituent of drain and oven-cleaning chemicals. The substance is corrosive and should be handled with caution.

Sodium hypochlorite *See*: Hypochlorite.

Sodium metasilicate A mildly alkaline cleaning agent commonly used in dishwashing formulations.

Sodium nitrate and nitrite *See*: Curing, Nitrite, Nitrate.

SOFHT (Society of Food Hygiene Technology) An organisation based in Lymington, England involved with food hygiene and training in the food manufacturing sector.

Soft cheese Made by a process not involving pressing. Those varieties ripened with moulds or other micro-organisms should be stored under refrigeration. British law is complex and involves a two-tier system of temperature control. Cut cheeses may currently be kept at a maximum of 8 °C but after 1 April 1993 they must be kept at a maximum of 5 °C. Whole cheeses and those ripening may be stored at the higher figure both now and in the future. In practice normal refrigeration (1–4 °C) would satisfy all current legal requirements. *See*: Appendix 1: Food Law, Cheese.

Softening of water Water hardness caused by dissolved substances may be removed by several means including boiling or by treatment with chemicals. On a large scale, water may be run through an ion exchange column so that dissolved calcium and magnesium ions which cause hardness are exchanged for sodium ions which do not. *See*: Hardness of water.

Soft ice-cream Mobile ice-cream vendors occasionally neglect cleaning, pressure of work being a frequently stated reason for this, and bacteria build up to dangerous levels. A soft ice-cream machine switched off uncleaned and left overnight may contain large numbers of micro-organisms the following day. *See*: Growth requirements (bacterial).

Soil

- Soil is a complex mixture of solid material, water, air, living organisms and the products of decay.
- It is one of the main reservoirs of micro-organisms, containing vast numbers of bacteria, fungi, algae, protozoa and viruses (usually plant pathogens).
- It has been estimated that just one gram of the top few centimetres of typical garden soil contains about 12 000 000 bacteria and 120 000 fungi, with moulds greatly outnumbering yeasts.
- Even though the vast majority of these micro-organisms are beneficial (helping to improve soil fertility) many of them are potential pathogens

or spoilage agents. Soil therefore presents important potential hazards in a food preparation environment.

Soil micro-organisms

- Many human pathogens can survive in soil for varying periods of time, some for many years.
- Most are spore-forming bacteria such as *Bacillus anthracis* (anthrax), *Clostridium botulinum* (botulism), *Clostridium perfringens* (gas gangrene and food poisoning), and *Clostridium tetani* (tetanus).
- Even non-spore formers, such as *Salmonella* species can survive for considerable periods (weeks or months).
- Also present in soil are members of the following bacteria: *Acetobacter*, *Alcaligenes*, *Chromobacterium*, *Enterobacter*, *Escherichia*, *Flavobacterium*, *Micrococcus*, *Mycobacterium*, *Proteus*, *Pseudomonas*, *Streptococcus*, and *Streptomyces*.
- Representative fungi include species of *Aspergillus*, *Mucor*, *Penicillium*, *Trichoderma* and many others.

Practical points

1) Unwashed vegetables are almost certain to have plenty of soil on them, and if they are prepared on the same work surface as, for example, cooked meat, cross-contamination may easily occur.
2) Surfaces and items coming into contact with soil must be carefully cleaned and disinfected before their use with other food.
3) Separate vegetable storage and preparation areas are preferred to minimise cross-contamination.
4) Soil organisms can be introduced into a food preparation area if personal hygiene is poor, for example, if the hands are not washed properly after handling raw vegetables such as potatoes, root vegetables such as carrots, and salad produce.

Solanin A substance, toxic to some people, produced in poisonous quantities in potatoes which have been stored in the light or cold. Since chlorophyll is also produced in these conditions, its green colour is an indication that solanin is present. There is evidence that high levels of this substance cause nausea, headaches and neurological disturbances. *See*: Alkaloids.

Sorbic acid An anti-mould agent, working best in slightly acid conditions. It is used as a food additive in cheese, bread, jam and other manufactured foods.

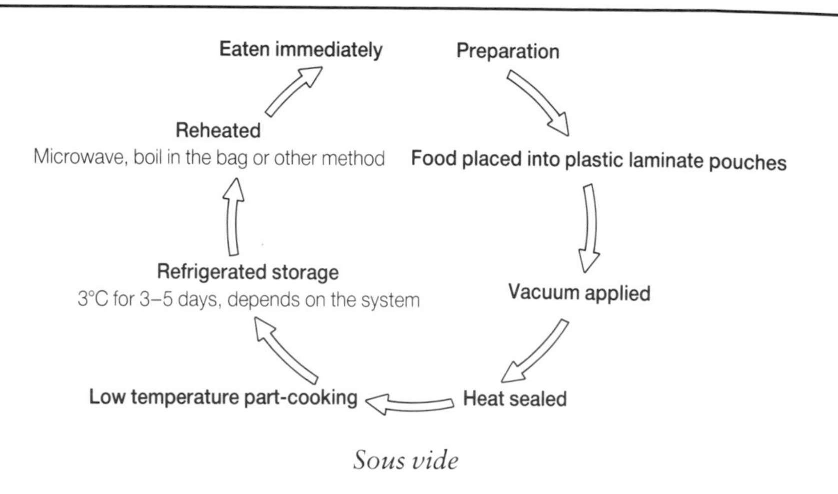

Sous vide

Sous vide A French term meaning 'under vacuum', used to describe an interrupted catering process in which lightly cooked food, often of high value (sometimes 'Cordon Bleu' style), is heat-sealed into plastic pouches in vacuum conditions.

- The product has a shelf-life under chill conditions (less than 5 °C) of a few days and is reheated when required.
- *Clostridium botulinum* favours both the vacuum and temperature found in this product so hygiene standards during production, storage and use are important.
- The process will not destroy spores or toxins.

Sparrows *See*: Birds.

Species In microbiology, a population of cells which are all essentially alike. For example, the bacterial genus *Clostridium* has a number of species, including *botulinum*, *perfringens* and *tetani*. (The species name is always written in the lower case.) Together with the generic name (in this case *Clostridium*), it forms the scientific name of the organism. Different species are thus named, and in this example we have *Clostridium botulinum*, *Clostridium perfringens* and *Clostridium tetani*.

Spiders More of a nuisance value than a health hazard although there is a risk of physical contamination of food with bodies.

Spirilla A group of rigid, spiral, corkscrew-shaped bacteria. They possess flagella.

Spirillum A curve-shaped bacterium. There are three forms: spirilla, spirochaetes and vibrio.

Spirochaetes A group of Gram-negative, slender, corkscrew-shaped bacteria. They differ from spirilla in that they lack flagella, but move by means of axial filaments. These are similar to flagella, but are contained under an external flexible sheath. Spirochaetes are found in contaminated water, in sewage, soil, decaying organic matter and in the bodies of humans and animals. Included in this group are the pathogenic *Treponema pallidum* (syphilis), *Borrelia* species (e.g. relapsing fever, transmitted by ticks or lice), and *Leptospira* species (leptospirosis, transmitted by contaminated water). Many spirochaetes are harmless saprophytes living in soil and stagnant water.

Spoilage of food

- The development of undesirable changes in food such as 'off' flavours and tastes, or changes in appearance such as colour, consistency or the production of gas.
- It is the result of natural decomposition and/or the effects of microbial growth in or on the food.
- Causes of food spoilage include bacteria, moulds, yeasts, and chemical reactions such as those leading to rancidity.
- Food poisoning bacteria such as salmonellae do not cause easily noticed changes in foods and do not make foods go 'off'.
- Proper, careful storage (e.g. refrigeration or in dry stores) will delay, but not prevent these changes.

Sporadic case An isolated or 'one-off' case of food poisoning not apparently connected with other cases.

Sporangiophore A structure produced by a mature fungal mycelium as part of the reproductive process. It is a stalked structure (an aerial hypha), supporting a spore case (sporangium).

Sporangiospore An asexual spore produced within a sporangium; they are produced as a means of reproduction in some moulds such as the common black bread mould, *Rhizopus*.

Sporangium A sac containing fungal spores. Typically each sporangium contains hundreds of sporangiospores.

Spore (bacterial) *See*: Endospore.

Spore (fungal)

- An asexual or sexual reproductive cell, the production of which is how fungi reproduce and spread. This is in stark contrast to the function of bacterial spores which are produced by some forms only, and as a means of survival.

- Millions of spores are released into the atmosphere and they are rapidly distributed by air currents.
- Many, once inhaled, may cause an allergic reaction, producing hay-fever type symptoms.
- Spores are of key importance in the identification and classification of fungi.
- Fungi produce sexual spores less frequently than they do asexual spores. They are produced most often under special circumstances. *See*: Ascospore, Basidiospore, Zygospore.

Spray drier A device consisting of a large chamber into the top of which a liquid is sprayed as a fine aerosol. The liquid droplets fall through a rising current of heated air and are dried. A typical product is dried milk. If the feed liquid is high-risk, such as milk, it may require pre-treatment such as pasteurisation.

Springer A stage in the deterioration and spoilage of canned foods in which thumb pressure on one end of the can causes the other to spring out. *See*: Blown cans, Canning.

Staining of bacteria Bacteria can be seen under the light microscope but they appear colourless. When stained they are much more clearly visible. A number of stains are commonly employed in the laboratory, including:

1) Gram-staining, which is a counterstaining method, and classifies all bacteria into one or other of two groups
2) staining of capsules and
3) the staining of endospores.

A great deal of information can be gained by looking at the stained and magnified organisms. For example, when attempting to identify a particular organism. However, the presence of a particular organism can only be confirmed by biochemical tests.

Stainless steel A durable, impervious and readily cleanable but expensive metal alloy with wide use in food rooms. There is a specific food grade containing 18% chromium and 8% nickel and known as 18/8. Care must be taken when cleaning this substance since it may be permanently damaged by the use of steel wool or scouring powers. *See*: Design of equipment in food premises.

Stalls (market) These are subject to similar hygiene regulations as other food premises. *See*: Appendix 1: Food Law.

Standing time (in microwave cooking) A period of time after initial cooking, during which heat spreads throughout the food (by conduction) and cooking continues. It is an important part of the cooking process. If it is not allowed for, the food will be undercooked.

Staphylococcus Any aggregate of spherically-shaped bacterial cells, that exist in random clusters after cell division. They are often described as being 'like a bunch of grapes'.

Staphylococcus aureus

- A bacterium which is responsible for many outbreaks of food poisoning. It is found in large numbers in cuts, boils, etc and it is also found in and around the nose and throat.
- It grows between 7 and 45 °C producing a toxin in the food (exotoxin) which causes food poisoning.
- It is a non-spore former, but the toxin is heat-resistant, being destroyed only gradually during boiling for 30 minutes.
- The onset period for the disease is 1 to 6 hours, and the illness lasts from 6 to 24 hours.
- Symptoms appear rapidly and are typically vomiting, subnormal temperature, abdominal pain and cramps, sometimes followed by collapse; death is rare.
- Because of these often mild symptoms, it is thought that many cases are not reported.
- It is a Gram-positive, facultative anaerobe, but grows best in the presence of oxygen.

Control

- Contamination is most often from food handlers, since the organism is found on the skin of about 40% of healthy people; good personal hygiene is vital.
- The toxin may not be destroyed by reheating or light cooking; thorough heating is necessary.
- Foods commonly involved are dishes which have been handled and then kept at room temperature before eating, such as cold meats, trifles, cream and custard-filled cakes. Refrigerated storage before use is important.
- It is the most salt and sugar resistant of the food poisoning organisms and so can cause problems with cured products such as ham and bacon. People sometimes falsely assume these are safe because they are salted.

- Similarly, the practice of salting butchers' blocks does not kill all staphylococci.
- Custard and cream products favour the growth of *Staphylococcus aureus* since the high-sugar content stops the growth of other bacteria (those less osmotically resistant).
- Refrigeration of susceptible foods is the easiest and most reliable method of preventing growth of this bacterium.

See: Exclusion of food handlers, Personal hygiene.

Starlings Sometimes called the 'flying rat' by pest control operators. *See*: Birds.

Star symbols Freezers and the frozen food compartments of some refrigerators carry a star symbol. Each star is 'worth' – 6 °C. This gives an indication of how long a frozen food may be stored and remain in good condition. The symbols are:

* up to one week (−6 °C)

* * up to one month (−12 °C)

* * * up to three months (−18 °C)

* * * * three months or longer (−18 to −24 °C)

Note: The 4-star rated freezer is capable of freezing fresh food.

Stationary phase *See*: Bacterial growth curve.

Statutory instruments Regulations and orders made under Acts of Parliament and having legal status. *See*: Appendix 1: Food Law, Codes of Practice.

Steam disinfection Steam may be used by trained persons during cleaning in the form of jets directed from a lance which is attached to a steam generator. It may also incorporate detergent metering. Accumulations which are difficult to remove and inaccessible areas are often steam-cleaned. Deep cleaning of ducting leading from extraction hoods may also use steam lances. *See*: Cleaning.

Steam fly *See*: German cockroach.

Sterile Free from any living thing.

Sterilisation

1) Any process that destroys (or removes) all living organisms

(including bacterial endospores) in or on an object. It very often involves heat and/or chemicals.

2) In milk processing, a temperature of 110 °C for about 30 minutes achieves commercial sterility. Other products such as ice cream and cream may also be sterilised.

The term is often misused and confused with disinfection. Sterilisation means 100% destruction of all living organisms, including their spores and it is very difficult to achieve. Disinfection (reduction of micro-organisms to a safe level) is more likely in processes such as cleaning. *See*: Pasteurisation, Turbidity test.

Sterilising sinks In washing up, a method of disinfecting crockery and cutlery, etc by standing them in hot water (82–88 °C) for at least 30 seconds. The term 'sterilising' is incorrect but is in common use. *See*: Washing up.

Stews, stocks and gravies High-risk products which must be subject to the same strict temperature controls as meat products.

Sticky boards Adhesive-coated traps used in pest control for insects and rodents in small infestations, or for survey purposes.

Stock rotation A system in which older stock is used first thus minimising risks of pest infestation and stock deterioration. *See*: 'Best-before' date, 'Shelf-life', 'Use-by' date.

Storage of foods Correct storage conditions for foodstuffs are essential to ensure freshness, maximum shelf-life, prevention of food poisoning and spoilage and prevention of pest infestations.

Elements of good storage

- Thoroughly check all goods inward to ensure that only good quality items are accepted.
- Damaged goods or those delivered at unacceptable temperatures (for example, partly thawed frozen foods) should be rejected.
- Each type of food has optimum storage conditions, for example:
  raw meat and poultry is best kept at around 0 °C
  meat products at less than 8 °C
  canned foods and many vegetables in dry conditions at 10–15 °C
  dried foods such as flour in tightly closed containers at 10–15 °C.
- In storage areas, construction and design should be such that proper cleaning is possible with foods stored off the ground.

See: Appendix 1: Food Law, Dry goods stores, Refrigerators.

Streptococcus Any group of spherically-shaped bacterial cells that remain attached in chains after cell division.

***Streptococcus* species** These are non-sporing Gram-positive bacteria, four groups of which are important in foods.

1) **Enterococcus** An example being *Streptococcus faecalis*, sometimes found in raw foods. It is unusual in that:
   a) it is thermoduric and likely to survive pasteurisation.
   b) it is tolerant of high salt concentrations and can grow on bacon and cause spoilage of canned ham.
   c) it can grow in alkaline conditions.
   d) it has a wide temperature range for growth (5 to 50 °C).

It is found in the intestinal tract of humans and animals, and is sometimes used as an indicator of faecal contamination of foods. *Streptococcus faecium* (more commonly from plant sources) is a very similar organism with similar properties. Evidence for food poisoning (associated with meat and poultry) is circumstantial.

2) **Lactic acid forming** Examples are *Streptococcus cremoris* and *Streptococcus lactis*, which are used as starters for hard cheeses. The latter has been shown to cause spoilage of raw milk, causing souring and ropiness. These organisms are not very salt tolerant.

3) **Pyogenic (pus forming)** Examples are *Streptococcus agalactiae* (mastitis in cows) and *Streptococcus pyogenes* (septic sore throat and scarlet fever), both of which have been found in raw milk. *S. pyogenes* has also been found in foods containing eggs or milk.

4) **Viridans** Examples are *Streptococcus thermophilus* (used as a starter for 'continental' cheeses and yoghurt) which may cause spoilage of canned foods and *Streptococcus bovis*, found in cow manure and saliva. Both forms are thermoduric.

Streptomyces A group of Gram-positive aerobic bacteria, commonly isolated from soil, which can cause undesirable flavours and appearance when growing on foods. They produce a musty or earthy taste, which may be absorbed by nearby foods. (Many species of *Streptomyces* have been used extensively for the production of antibiotics and more recently for the industrial production of enzymes.)

***Strongylus* species (Nematodes)** Small worms affecting internal organs of food animals, which if badly affected, should be condemned.

Stuffing Should be cooked separately from poultry since filling the body cavity makes cooking times longer and increases the risk of organisms such as salmonellae surviving. It is possible that the stuffing will become contaminated and thus become more dangerous than the poultry itself. *See*: Centre temperature.

Sublimation The direct change from solid (ice) to water vapour without liquid forming. *See*: Accelerated Freeze Drying, Freezer burn.

Sucrose Table sugar.

Sugar A preservative which causes the removal of water from cells, including bacterial cells, by osmosis. Moulds are less affected. Foods with a high sugar content such as jam, confectionery and some bakery items are unlikely to cause food poisoning. *See*: Available water, Osmosis.

Sulphiding A chemical reaction between sulphur-containing amino acids in meats and some vegetables (e.g. peas) and iron in cans produces iron II (ferrous) sulphide. This is a blackish-coloured substance which may stain food in cans, or the inside of cans especially along side seams. It may safely be trimmed off corned beef and simlar foods.

Sulphur dioxide An acidic gas which is the active component of a number of preservative compounds. It is very widely used in beers, wines, meat products and many other foods. It is toxic to most animals and levels in food are controlled.

Sun-drying An ancient method of preservation, used in Mediterranean regions for thousands of years to produce raisins, sultanas other dried fruits and dried meats/fish. The technique is still in use today, problems being attack by birds and pests. Food poisoning bacteria will not live on dried foods. *See*: Dehydration.

Surfactants (Surface-active agents) *See*: Detergents.

Swine erysipelas An infectious bacterial disease of pigs causing skin discolouration and septicaemia which may affect humans and other animals.

Sycosis barbae A staphylococcal-caused rash, occurring in the beard region of the face. It results in the formation of many pus-containing spots, which, if allowed to contaminate food (e.g. by the fingers), could cause staphylococcal food poisoning.

Symbiosis A close physical relationship between organisms of two different species that may or may not be beneficial to the organisms involved. *See*: Commensalism, Mutualism, Parasitism.

Symptomless excreter Some people who harbour disease-causing organisms inside their bodies sometimes excrete them, which may then contaminate foods, surfaces, etc. They themselves never show any symptoms of any disease and are of great importance in the food industry since such people can easily be responsible for causing outbreaks of bacterial food poisoning, without being aware of it. As they show no symptoms, they would not be excluded from work, as would someone suffering from diarrhoea for example. *See*: Carrier, Typhoid Mary.

Symptoms of food poisoning

- Nausea
- vomiting
- diarrhoea
- abdominal pain and cramps.

These may vary widely from symptoms so mild that they are ignored, to those which are so severe that affected individuals suffer collapse and/or severe discomfort. There may also be additional symptoms, such as fever or chill, headache, aching of limbs, and in some cases (such as in botulism) paralysis and death are not uncommon.

Synergism When two kinds of micro-organisms or chemicals act together, they may be able to bring about changes (such as fermentations) that neither could produce alone.

Taeniasis The disease caused by parasitic tapeworms, particularly the Cestoda.

Taenia hydatigena A tapeworm whose intermediate host is cattle and other food animals, the final host being dogs.

Taint in meat Undesirable odours and tastes, resulting from the growth of bacteria on the surface of meat, often evident before other signs of spoilage.

Tasting of food If food is to be tasted during cooking, for example, to check seasoning, it is important that a clean implement is used and it should not be re-used without prior washing. Once used it is certain to be contaminated with saliva from the mouth, which will contain bacteria, including *Staphylococcus aureus*. The re-use of spoons without washing in between has been found to be responsible for outbreaks of food poisoning. Similar dangers of probable contamination of food are likely if a finger is used in order to taste the food. *See*: Hands, Mucous membrane, Personal hygiene.

Tap proportioner A metering device connected to a water tap which controls the amount of liquid substances such as detergents which require mixing with the water.

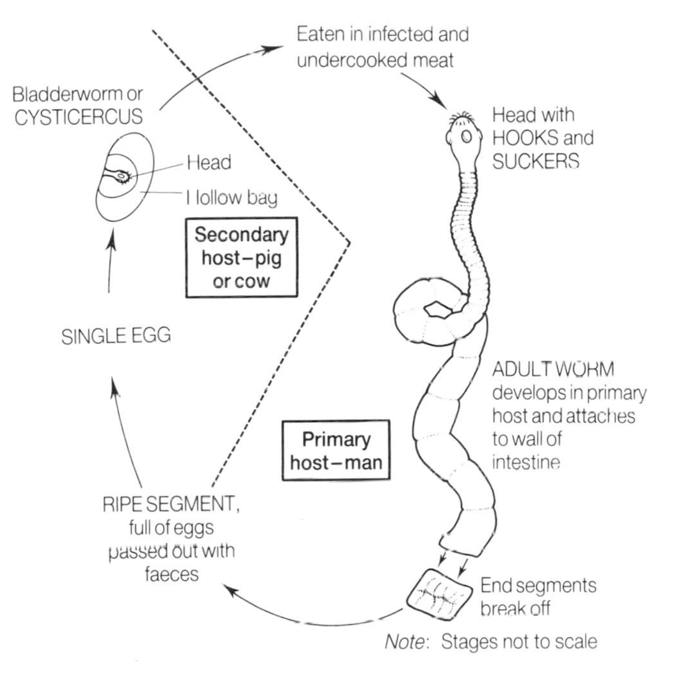

Life cycle of the tapeworm

Tapeworms Parasitic worms having a small head and a long body consisting of many segments (proglottids). The head possesses suckers (some also have hooks) which enable the worm to attach itself to the lining of the intestines.

Segments are continually produced at the neck region as long as the worm is alive and attached. Some grow up to 15–20 metres in length and may live for up to 25 years.

The worms are hermaphrodite (each segment containing a set of male and female reproductive organs) and segments furthest from the head contain fertilised eggs which pass out with the faeces of the host.

Several species exist which affect food animals, domestic animals and others. Examples include *Diphyllobothrium* (fish), *Taenia* (beef and pork), *Echinococcus* (dog, sheep, cattle and horses) and *Toxocara* (many animals including dogs, cats and foxes).

The beef tapeworm is typical of the risks, and necessary precautions that should be taken for most of these parasites. Generally, symptoms are not severe but vomiting, anaemia, diarrhoea and loss of appetite may occur.

Temperature Several temperatures are important in food hygiene from freezer temperature at −18 °C to the process temperature of 132 °C for UHT milk. *See*: Danger zone.

°C	
180	Hot cooking oil
132	UHT milk (one second)
121	Fo – *Cl. botulinum* killed in 3 minutes
100	Water boils. Bacteria killed more quickly
88	Final rinse in washing up
82	Hot cupboard
72	Pasteurisation (milk)
70	Minimum centre temperature for cooked foods
63	Top of danger zone. Bacteria killed slowly
55	Main wash water in washing up
37	Human body temperature
20	Comfortable room temperature
15	Dry goods store
10	Bacteria start to become dormant
8	Lower danger zone for some high-risk foods
5	Lower danger zone for higher risk foods
4–1	Refrigerator
0	Water freezes
−6	One star freezer
−12	Two star freezer. Some ice-cream freezers
−18 to −25	Deep freeze storage
−35 and below	Blast freezer. Very low temperature storage

Temperature zones

Temperature control

- The practice of keeping high-risk food outside the range of temperatures in which food poisoning and food spoilage bacteria grow.
- This is 5–63 °C and is called the 'danger zone'.
- Refrigeration and cooking are the two most common methods of controlling temperature.
- A commonly quoted motto in food hygiene is:

'Keep it hot, keep it cold or don't keep it at all'.

See: Appendix 1: Food Law, Bacterial growth requirements, Mesophiles.

***Tenebrio* species** Large, oval, dark-coloured beetles which include *T. obscuris* – the dark mealworm and – *T. molitor* – the yellow mealworm. Adults, which may grow to 3 cm, and larvae affect mainly cereals and thrive in damp dark conditions. Their presence is an indication of long-standing cleaning problems since infestations take many months to develop.

Terrapin Many pets, including such 'exotic' types, are known carriers of *Salmonella*.

Tetra-Pak™ A laminated paper container developed for packaging UHT milk and other liquids such as fruit juices. The container is made by machine from a roll of laminate while the liquid is being filled under aseptic (sterile) conditions.

Thawing

- If food is to be cooked, adequate thawing is vital to ensure that the heat of cooking can kill any pathogens and is not wasted in melting ice.
- Thawing in a refrigerator at less than 5 °C may take many hours but will limit growth of bacteria. It is probably the safest method.
- Thawing at room temperature results in a large increase in numbers of bacteria on the surface while the inside may remain frozen. If cooking is thorough, this may not be a problem although some bacteria may produce toxins which are not destroyed by cooking.
- Controlled thawing at 10–15 °C, keeping food separate and taking care that all equipment used is disinfected and that juices do not cause cross-contamination, are important factors.
- Microwave thawing has the disadvantage of producing localised hot or cold spots. It must be done on the 'defrost'setting following the manufacturers' recommendations and can be very effective.

- Certain small-sized foods such as vegetables and beefburgers do not require thawing and may be cooked from frozen. This is not true for large items such as frozen poultry.

Thermal death rate curve A graph of the log of the number of surviving cells against time of heating at a constant temperature. From this graph the thermal death time curve is derived.

Thermal death time The length of time required to kill all bacterial cells of a given species at a given temperature.

Thermal death time curve A graph of time against temperature showing equivalent time/temperature combinations lethal to an organism. For example *Cl. botulinum* may be destroyed by heating at 100 °C for 6 hours or at 121 °C for 3 minutes. *See*: Fo value.

Thermoduric Heat resistant. Such micro-organisms endure high temperatures, waiting for more favourable conditions to develop. Examples include the *Bacillus* and *Clostridium* groups of bacteria. *See*: Endospore.

Thermometers Probe thermometers are now commonly used in many food premises.They should be:

1) calibrated before use (iced water at 0 °C and boiling water at 100 °C are useful tests) to check accuracy.
2) disinfected in between uses to avoid cross-contamination, a disinfectant wipe is often used both before and after taking a reading
3) inserted at the correct depth to ensure that it is the temperature at the centre of the food which is being measured.

In refrigerators and freezers, a fixed thermometer may be fitted by the manufacturer or a small strip thermometer used.

Temperatures of refrigerators (1–4 °C) and freezers (−18 °C) should be checked and recorded several times per day. Some larger installations use chart recorders or computer-controlled systems fitted with alarms. *See*: Due diligence, Refrigerators.

Thermophile An organism whose optimum temperature for growth is above 45 °C. *Bacillus stearothermophilus* is such an organism which sometimes causes spoilage in canned foods.

Tiling Ceramic or other impervious tiles with a suitable waterproof grout are excellent surface materials for food room use including walls, floors and partitions. They are durable if the correct cleaning agents are used but initially expensive. The main disadvantage is that they are brittle and may crack on impact, becoming unsatisfactory since the crack may harbour bacteria and food scraps. Cracked tiles are impossible to clean properly.

Tin

1) A corrosion-resistant metal used to coat the mild steel plate from which cans are made. The common name for cans is tins.
2) Food poisoning has been reported from opened cans of fruit which have become contaminated with tin dissolved from the can by the action of fruit acids. It is therefore advisable to decant the contents of part-used cans into suitable covered containers.

TMA Some types of fish produce the chemical substance trimethylamine during decomposition. The amount may be used as an indication of freshness. *See*: ATP.

Toadstools A popular, non-scientific term used to describe large fungi other than the common edible mushroom. Toadstools are generally thought of as being poisonous.

Toilets *See*: Sanitary conveniences.

Torry kiln A mechanical trolley-loaded kiln used in the smoking of fish.

Towels (hand, paper, roller) *See*: Drying of hands.

Toxic Poisonous.

Toxin A poison of bacterial or other origin. Ingestion of toxins in foods is a common cause of food pisoning. *See*: Endotoxin, enterotoxin, exotoxin, haemagglutin, lectin, Paralytic shellfish poisoning.

Toxocara Tapeworms which are common parasites of dogs (*T. canis*) and cats (*T. felis*) producing larvae which may cause serious damage to various organs in humans. Infection is thought to be by contact with animal faeces, the urban fox being one reservoir of this organism. Other aminals may also be affected.

Training of food handlers

- Persons handling food should possess sufficient knowledge of food hygiene to enable them to act in a safe and hygienic manner.
- A trained workforce is likely to be more confident and efficient, less likely to cause food poisoning or other problems concerning the product and be better motivated.
- It is unlikely that due diligence could be proved if a food business was operating with untrained staff and problems concerning food safety occurred.

- Various types of training are appropriate, depending on levels of responsibility and work done.

Basic courses (about 6 hours)
- For all food handlers such as cooks chefs, bar staff, etc.

Intermediate courses (about 18 hours)
- For supervisors and some management.

Advanced courses (about 36 hours)
- For managers and trainers.

See: Due diligence, IEHO, RSH, RIPHH, SOFHT.

Transient organisms Micro-organisms that are present on an animal, including humans, for a short time without causing disease. *See*: Skin.

Transmissible disease *See*: Communicable disease.

Traps Insects, birds and rodents may be physically controlled by catching them, normally in a baited trap. Such devices need expert location and there should be no alternative foods available if the bait is to be successful. There is a problem in disposing of live or dead bodies. Traps may also be used in surveys to estimate the extent of an infestation. They are an advantage in that poisonous baits are not left uncovered in food premises and the captured pest can be removed rather than being allowed to go off and die, possibly in or on a food product. *See*: Flypapers, Sticky boards, Ultra violet radiation.

Traveller's diarrhoea Common name for food poisoning acquired during travel and caused by a number of organisms, such as *E. coli*.

Trematodes Flatworm parasites of animals and humans, the most common being the liver fluke (*Fasciola hepatica*).

Tribolium The flour beetles including *T. castaneum* (rust red flour beetle),*T. confusum* (confused flour beetle) and *T. destructor* (dark flour beetle). They are common pests of stored cereals, especially flour. Adults vary from 3–6 mm in length and are brown in colour, with *T. destructor* sometimes being black. *See*: Beetles.

Trichinella spiralis
- The parasitic trichina worm which produces cysts in humans, rats and pigs.
- Eating pork containing viable cysts results in small worms (up to 4 cm long) maturing in the intestines.

- Fertilised female worms burrow into the intestinal lining and deposit large numbers (thousands) of larvae which enter the bloodstream and are carried throughout the body.
- The larvae form cysts in muscles including the diaphragm and tongue and may cause severe problems. After some time the cysts in humans calcify and die.
- The reservoir of infection is the pig which contracts the disease from eating undercooked food containing infected pork or from rats.
- Pork may be made safe by thorough cooking or freezing.

Trichiniasis or trichinosis The disease in humans or animals caused by the parasitic worm *Trichinella spiralis*.

Tuberculosis In humans, a chronic communicable disease of the respiratory system caused by the Gram-positive bacterium *Mycobacterium tuberculosis*. It may be contracted from food such as milk or by droplet infection. There are two forms: human and bovine. Bovine tubercuosis has been eradicted by a control programme in some countries including Great Britain, but it is rife in many others. Heat treatment of milk destroys the bacterium. *See*: Milk, Pasteurisation.

Turbidity test A post-processing test for sterilised milk depending on soluble proteins being denatured by heating above 100 °C. A positive test indicates insufficient heating or the presence of raw milk.

Turkeys Share common food hygiene problems with chickens but the large size of some birds makes thawing, cooking, cooling and temperature control in general more difficult. *See*: Poultry.

Two-sink system (Double sink system) A manual method of washing up using:

1) a sink containing hot water at 55 °C with a detergent and
2) a second sink with water at 82–88 °C or with cooler water containing a chemical disinfectant.

Less soiled items such as glasses are washed first. The sequence of washing up is:

a) Pre-clean or scrape items to remove large deposits.
b) Wash in the first sink to remove grease and soil.
c) Rinse in the second sink to remove residues and to heat up and disinfect cleaned items.
d) Air-dry the hot items.
e) Final storage away from possible contamination.

See: Cleaning, Dish-washing, Soil.

Typhoid fever

- One of a number of food-borne diseases affecting large numbers of people wordwide, principally from contaminated water supplies.
- The organism responsible is *Salmonella typhi* which produces a severe, sometimes fatal fever (up to 20% mortality if untreated).
- General symptoms including headache, pains, constipation and diarrhoea.
- The organism affects many organs in the body any of which may become inflamed.
- Outbreaks often occur when water and sewage systems are inadequate or damaged and the disease has also been spread by milk and other foods.

See: Aberdeen typhoid outbreak, Milk, Paratyphoid fever.

Typhoid Mary Mary Mallon, who worked as a cook in New York State in the early part of this century. She was a carrier of typhoid fever and was responsible for several outbreaks of the disease, which resulted in three deaths. Her case became well known through the attempts of the State to restrain her from working in the food trade, which included periods of imprisonment.

Tyrosine An amino acid causing aesthetically unacceptable white spots in the livers of sheep.

UHT (Ultra Heat Treated) A preservation process developed for milk in which commercial sterility is achieved by heating to 132 °C for one second followed by aseptic filling into sterile containers, normally made from plastic laminated board. The product has a shelf-life of six months without refrigeration.

Ultra sonic sound Attempts have been made to control various pests with emissions of high-frequency sound above the range of human hearing. The technique works for a time but acclimatisation then occurs.

Ultra violet radiation (UV) A type of radiation used in electric flying insect killers to attract these pests. It also has limited use for sterilisation purposes, e.g. the surfaces of fruits, and for small quantities of water. *See*: Electromagnetic radiation.

Undulant fever *See*: *Brucella abortus*.

Unfit food Food which is not suitable for human consumption. The possession of such food in a food business (unless in refuse containers) may be an offence since all food is deemed to be on food premises for the purpose of human consumption. Unfitness may be judged by bacteriological standards or by organoleptic assessment including the presence of physical contaminants. Food found to be unfit, including other batches, may be seized and destroyed by authorised officers such as EHOs.

Untreated milk *See*: Raw milk.

Urinals One per 25 persons up to 100 persons, and then 1 per 40 persons thereafter is deemed sufficient (Sanitary Conveniences Regs 1964 – premises other than factories).

'Use-by' date A date reserved for labelling perishable high-risk foods which may cause a health risk due to growth of micro-organisms after a certain date. Typical foods include meat products and cook-chill foods. Labelling of these foods with a 'use-by' date is a legal requirement in Britain. *See*: 'Best-before' date, Food Labelling Regs, Stock rotation.

USFDA (United States Food and Drug Administration) The body in the USA roughly equivalent to MAFF and DoH in the UK in respect of some food matters.

Vacuum-packed foods Mainly meats and fish which are heat-sealed into laminated plastic or nylon pouches. The oxygen-free conditions may favour the growth of *Clostridium botulinum* and the process must be hygienically operated to prevent any product contamination. As soon as the pack is opened, the food should be treated as if it were fresh and may require refrigeration before use. *See*: Sous vide.

Vector A transmitter and reservoir of disease. They may be:

1) Inanimate (non-living), including food, water, saliva, pus and contaminated eating utensils
2) Animate (living), generally insects or animals, including humans.

See: Animals, Droplet infection, Personal hygiene, Rodents.

Vehicle of food poisoning

- The means by which micro-organisms may travel from place to place.
- This includes food itself, hands, wiping cloths, cutting boards, knives, etc.
- Each year the same categories of food appear in the statistics for reported cases of bacterial food poisoning. Meat, meat products and poultry are the largest category, whereas highly salted, sugary and acid foods are rarely responsible.

Vehicles (delivery) Such vehicles used for food must be constructed and maintained so that they are clean. Temperature control in refrigerated delivery vehicles is important and should be regularly checked. *See*: Cross-contamination, Delivery vehicles.

Vegetables

- Should be stored separately to keep soil bacteria away from other foods.
- Most store best in cool conditions (10–15 °C).
- They should be prepared in a separate area of the kitchen since there is a risk of cross-contamination from the soil on the surface onto other foods.
- Freshly cooked vegetables should be safe as long as they are eaten straight after cooking.
- Cooked vegetables, if not to be eaten immediately, should be refrigerated or kept hot.
- All salad vegetables should be washed thoroughly before consumption.

See: Appendix 1: Food Law, *Clostridium perfringens*.

Vegetative cell Cells which are actively growing as opposed to those which are reproducing or 'resting' (in the endospore state). It is during the so-called vegetative phase or state that exotoxins are produced and released.

Vending machines

- An increasing variety is available from simple pack vending machines to those which prepare drinks, including soups.
- Construction and design follows the basic principle of ability to be kept clean, with smooth impervious services internally and externally.
- The name and address of the operator must be clearly visible on the machine.
- If high-risk foods are stored then proper stock rotation and temperature control must be carried out.
- Refrigerated vending machines may go 'off sale' automatically if the temperature rises above the legal level.

Ventilation

- Depending on the use of a room, a certain number of air changes per hour are required so that heat, water vapour and fumes generated by people and equipment do not build up to unacceptable levels.
- Kitchens may require up to 20 air changes per hour.
- Systems range from simple hoods and ducting to complex air

conditioning systems. All of these must be included in the routine cleaning and maintenance programme.
- Filters should be regularly inspected otherwise they will lose efficiency and may become a fire hazard.
- In air conditioning systems, there is a risk of creating conditions favourable to the growth of the bacterium causing Legionnaire's disease.
- The design of ventilation systems is a specialist task.

Vermin Undesirable pests such as rodents, cockroaches, starlings, etc.

Vibrio

1) A term used in the lower case ('vibrio') to describe the shape of certain bacteria, which are curved or comma-shaped.
2) When written as '*Vibrio*', the name of a group (genus) of bacteria.
3) At least three vibrios are found in British waters (*V. parahaemolyticus*, *V. fluvialis* and *V. alginolyticus*).

Vibrio cholerae A slightly curved rod-shaped Gram-negative bacterium which is responsible for the disease cholera, due to the release of a powerful enterotoxin. The organism survives well outside the body, for example in cool, moist places it will survive for days. It is easily killed at high temperatures (10 minutes at 55 °C.) and is sensitive to chemical disinfectants. Cholera is most often caused by contaminated water supplies. Many outbreaks are caused by shellfish, e.g. Italy (1974) due to mussels and USA (1980s) due to crabs and shrimps.

Vibrio parahaemolyticus
- A slightly curved, rod-shaped, Gram-negative, halophilic bacterium, which causes food poisoning by infection.
- It usually inhabits coastal salt waters and is transmitted to humans mostly by shellfish.
- Symptoms include a burning sensation in the stomach and abdominal pain, together with vomiting and watery diarrhoea.
- The incubation period is less than 24 hours and the illness lasts for several days; the fatality rate is low.
- The organism is easily killed by proper cooking of the seafood, hence prevention of re-contamination of cooked shellfish by raw shellfish or by sea water is essential.
- It is a very common causal agent of food poisoning in Japan (45–70% of total reported cases).

Vibrio vulnificus A serious contaminant of seafood in the Atlantic and Pacific Oceans and the Gulf of Mexico, having around 50% fatalities. It has the nickname 'terror of the deep'. It produces septicaemia (blood poisoning) in its victims. It is not found in British waters.

Vinegar Contains acetic (ethanoic) acid and has a sufficiently low pH (less than 4.5) to prevent the growth of pathogens. Pickled foods however are subject to mould growth and preservatives such as benzoic acid are also incorporated in commercially produced pickles.

Vinegar fly (*Drosophila* species) Various fruit flies which may breed in fermenting materials such as overripe fruit, decaying vegetables, vinegar and brewing liquid. Refuse control and covering of food are the best control measures. *See*: Flies.

Virion A fully complete unit of a virus particle, usually composed of a nucleic acid core surrounded by a protein coat.

Virucide Literally means an agent used to 'kill' viruses. However, since there is some debate as to whether viruses are living organisms, it is safer to describe a virucide as an agent used to inactivate viruses. The application of heat is a simple but effective agent and some viruses are sensitive to chemical disinfectants.

Virulence The degree to which a pathogen can cause harm in a specific host, i.e. relative pathogenicity. It may be due to toxicity and/or aggressiveness (invasiveness). There is a range of virulence, extending from avirulent (harmless) to highly virulent (extremely harmful/deadly).

Viruses

- Cases of food poisoning caused by viruses were recorded as being over 15 000 in Britain in 1990, this being similar to the number of reported cases of salmonellosis.
- The major cause of these viral outbreaks is probably by the faecal-oral route although contaminated foods, water and shellfish have also been noted.
- In the UK the most common appear to be the small round unstructured viruses (SRSV) of the Norwalk type although Astrovirus and Calcivirus have been noted in seafood.
- Viral food poisoning normally develops within 48 hours and causes nausea, vomiting, cramps and diarrhoea; symptoms usually subside after 2 days.

Characteristics of viruses

- Very small and simple parasitic agents which must be inside living host cells in order to reproduce.

- Due to their non-cellular nature they are not considered to be living organisms.
- They mostly consist only of a core of nucleic acid (DNA or RNA) surrounded by a protein coat, but some possess an outer envelope as well; the single unit is called a virion.
- Their size is measured in nanometres (billionths of a metre) and they are only visible under the electron microscope.
- The number and variety of viruses is unknown as is the range of diseases they may cause.

Viruses are generally sub-divided into three main groups based on the kind of host they infect:

1) bacterial viruses (bacteriophages)
2) animal viruses
3) plant viruses.

Viruses are unable to multiply unless they are inside a living host cell, hence they cannot increase in numbers in foods. However, they can remain viable in food or water for variable periods of time. Following ingestion of the virus-contaminated food or water, the virion, after entering the host cell 'takes over' the cell totally to produce many copies of the virus particle. This often results in damage to or death of the host cell. Very small numbers of virions are required to cause disease. *See*: Enterovirus, Hepatitis A/B, Norwalk virus, Poliomyelitis.

Walls As part of the structure of food rooms, these should be constructed with a durable, impervious surface capable of being easily cleaned. A light colour is preferable so that cleanliness may be seen and also to assist in lighting. Several costly but effective wall finishes are ideal for food rooms, particularly if there is likely to be contact with food. Examples include:

1) ceramic tiling with a waterproof grout
2) plastic coated boarding
3) stainless steel
4) fibre glass and other resins
5) purpose-formed polypropylene sheeting.

Painted surfaces, using a special food grade finish are satisfactory as long as flaking does not occur. Washable wallpaper may be of limited use and normal domestic wall finishes should not be used in production areas. *See*: Design of food premises.

Warfarin An anticoagulant rodenticide developed in the 1960s. It causes rats and mice to die slowly so that others do not associate eating the poison with their death. *See*: Rats.

Wasps These can be a nuisance and also theoretically carry food poisoning organisms on their feet and bodies. From mid-summer they are attracted by sweet substances, such as fruit juice, sugar, syrups and preserves. A 3 mm mesh screen will prevent entry into premises.

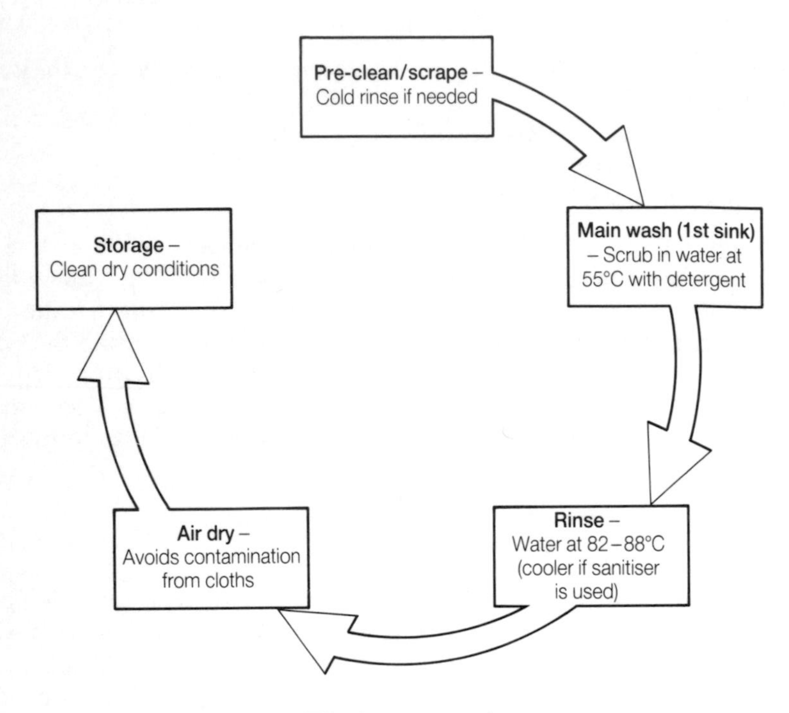

Washing-up cycle

Washing up Several systems ranging from manual to machine are in use, all being based on the following sequence of operations.

1) Pre-cleaning (scraping of plates, pans etc).
2) Main wash using detergent at around 55 °C. This removes greasy soil and is not so hot that food (such as egg) is baked onto surfaces.
3) A rinse in water at 82–88 °C. This removes traces of detergent, heats up items so that they may air dry and also disinfects them. The rinse may contain sanitising chemicals, particularly if the water temperature is too low.

4) Air-drying.
5) Correct storage away from contamination.

In manual washing up one sink may suffice, working on a fill and empty system, but two sinks are preferable. The first contains detergent and the second is the rinse sink. Automatic systems normally use similar cycles and it is important to use chemicals correctly and to monitor temperatures. *See*: Cleaning, Disinfectant, Sterilisation, Two sink system.

Washing facilities Suitable washing facilities for food handlers must be provided in food premises so that the required standard of personal hygiene may be achieved. As a rough guide, one wash hand basin per 15 persons is required; the exact numbers are legally determined, but one per 10 persons is more desirable, particularly in high-use areas. *See*: Hand washing, Personal hygiene.

Wash hand basins Separate wash hand basins, not used for any other purpose, must be provided in food premises. They must have soap, hot water, suitable drying facilities and a nailbrush. *See*: Basins, Hand washing, Personal hygiene, Washing facilities.

Waste (control and disposal of) In food premises, a suitable number of containers with close-fitting lids must be provided for wet and dry waste. Plastic sack holders or waste bins are often used and must be periodically emptied to separate refuse areas for the wet and dry waste. These must be kept clean to discourage animals such as dogs, flies and rats. Collection from these areas must be regularly carried out. It is important that waste is covered and contained at all stages and that all containers are kept in good condition. *See*: Compaction.

Water activity *See*: Available water (a_w).

Warehouse moth (*Ephestia elutella*) A very common pest of stored products including cereals, chocolate dried fruit and nuts. *See*: Moths.

Water

1) A food business cannot operate without an adequate supply of wholesome water. It is the responsibility of the supplier to ensure suitable quality in the water delivered to a building.Contaminated water may affect food.

2) Water may be classed as being food, for example a glass of water given to a customer in a restaurant, or ice-cubes given in a drink.

See: Aberdeen typhoid outbreak, Food-borne disease, Infections (water-borne), Personal hygiene, Washing facilities.

Waterproof dressing

- It is a legal requirement in the food industry to cover skin lesions such as cuts with suitable dressings.
- They not only prevent the lesions from becoming infected, but they also prevent the escape of blood and/or pus from the wound, both of which may contain bacteria, such as *Staphylococcus aureus*, which may then contaminate food.
- Special bright blue coloured versions are available which are easily seen should they get into food.
- In food manufacturing, metal strips may be incorporated into the dressing for magnetic detection.

Water-borne infections *See*: Infections (water-borne).

Weedy fish Foul smelling iodine-like smell caused by fish eating various marine organisms.

Weevils Small beetles (up to 5 mm in length) having a characteristic snout.

Weil's disease

- A disease caused by a parasite of rats (*Leptospira icterohaemorrhagiae*, a spirochaete); it is commonly present in their kidneys and urine, and is apparently not harmful to them.
- It also occurs in dogs, in which it is pathogenic, in pigs and other rodents such as field mice.
- The parasite may enter the human body through the skin, mucous membranes or skin abrasions, following contact with water contaminated by the urine of infected rodents.
- It is possible that infection may follow ingestion of food or water contaminated by infected rodent urine and cases of infection have followed rat bites.
- The use of stagnant water for drinking, preparing food and for bathing or recreation (e.g. windsurfing) should be avoided.
- The disease has been frequently reported among workers at slaughterhouses and refuse tips after contact with items contaminated with rat urine.

Windows As part of the structure of food rooms these must be constructed from durable and impervious materials capable of being effectively cleaned. Screens and nets may be attached as part of proofing against the entry of flying insects and birds. Windows may also be a source of heat (solar gain) and their position may be of importance when considering the location of equipment such as refrigerators.

Wood

- Potentially a problem material in food premises due to its absorbent properties, splintering and difficulty in cleaning.
- Bare wood has no place in food rooms with the possible exception of cutting boards for raw meat; these must be constructed of high density wood and be capable of being cleaned.
- To date, a completely acceptable substitute for wood has yet to be found.
- The practice of salting wooden butchers' blocks is of some use, but it does not destroy toxins, spores or staphylococci.
- There is a school of thought which considers that wooden chopping boards are illegal for food use since absorbent surfaces are not allowed under the Food Hygiene (General) Regulations 1970. *See*: Chopping boards.

Work-flow *See*: Linear flow.

Worktops *See*: Food-contact surfaces.

WHO (World Health Organisation) Based in Geneva, Switzerland and having interests in communicable diseases of all kinds, including food poisoning and many other health matters.

Wrapping materials Materials which come into contact with food should be inert and not cause any contamination. Newspaper may not be directly used to wrap foods with the exception of, for example, raw vegetables and unplucked poultry. *See*: Aluminium, Cling film, Packaging.

Worms Many parasitic worms which are harmful to humans and animals may be found in food. These include *Ascaris lumbricoides*, Nematodes, Roundworms, *Taenia* species, and Trematodes.

***Xenopsylla cheopsis* (the rat flea)** It is responsible for the transmission of 'plague', known in the Middle Ages as the Black Death.

Xerophile A micro-organism capable of growing under 'dry' conditions (Xero = dry, -phile = lover). Xerophilic fungi are able to grow under relatively dry conditions, i.e. with an a_w of 0.65 to 0.75. It is a general property of moulds that they can grow under conditions that are too dry for for the majority of yeasts and bacteria. Examples of xerophilic moulds include *Aspergillus echinulatus* (a_w around 0.65) and *Xeromyces bisporus* (aw around 0.6). *Note*: A dehydrated food commonly has an a_w of less than 0.6.

Yeasts

- Single-celled fungi, widely distributed in nature, and frequently seen as a white powdery coating on fruits and leaves.
- Some are capable of the fermentation of carbohydrates to produce alcohols and carbon dioxide.
- Carbon dioxide is used in baking for raising bread, while alcohol production is the basis of the brewing and wine-making industries; in both industries species of *Saccharomyces* are widely used.
- Wild types of yeasts are responsible for the spoilage of many types of food such as meat, wine, beer, and fruit juice.
- Most common yeasts grow best with a plentiful supply of water but need less moisture to grow compared to bacteria.
- The temperature range for growth is similar to moulds, which is an optimum of 25 to 30 °C, and a maximum of 35 to 47 °C; some can grow at 0 °C or less.
- They prefer an acid environment, around pH 4 to 4.5.

Yellow swarming fly (*Thaumatomia notata*) A small insect, 3 mm in length, occasionally invading dwellings and food premises at the start of cold weather. *See*: Flies.

Yersinia enterocolitica A bacterium capable of growing at refrigerator temperatures which causes the food-borne illness known as yersiniosis. Foods involved are commonly pork and other meats, raw milk or any contaminated raw or left-over food. Thorough cooking, protection from contamination and rodent control are effective control measures. Symptoms produced include acute intestinal upset, with fever, vomiting and diarrhoea.

Yoghurt A milk product formed by the action of bacteria such as *Lactobacillus bulgaricus* and *Streptococcus thermophilus*. These consume lactose (milk sugar) and produce lactic acid which causes milk protein to clot (denature). If the pH is below 4.5, yoghurt is not a high-risk food. If above this figure it must be stored in a refrigerator before use. *See*: Appendix 1: Food Law.

Young persons (cleaning of dangerous machines) Persons under the age of 18 years may not clean certain machines such as meat slicers, chippers and some mixers which are legally defined as being dangerous. Any person cleaning such machines must first be trained.

Zinc One of a number of metals which can cause food poisoning. It may enter food by being dissolved by fruit acids. *See*: Galvanised steel.

Zone of maximum ice crystal formation The speed of freezing determines the quality of the final product since the less time food spends in the temperature range 0 to −4 °C the smaller the ice crystals which are formed. Modern quick freezing rates are around 3–10 cm/hour (movement of ice front) in blast or fluidised bed freezers, whereas slow freezing would occur at about 0.2 cm/hour. *See*: Frozen food.

Zoonoses Diseases that occur primarily in wild and domesticated animals, but they can be transmitted to humans. Transmission can occur by many routes:

1) direct contact with infected animals
2) contact with hides, fur or feathers
3) by insects
4) contamination of food and water
5) consumption of infected animal products.

Many such diseases occur including anthrax, foot and mouth, glanders, salmonellosis, swine erysipelas, taeniasis, trichinosis, tuberculosis, undulant fever and vaccinia.

Zygospore A large sexual fungal spore, produced as a means of reproduction. It is enclosed in a thick resistant wall. Such spores are formed by *Mucor*, a common spoilage mould.

Appendix 1: Food Law in Britain

This section is not intended to be a comprehensive guide to the wide-ranging and complex subject of food legislation, but is meant as a guide to the more important issues affecting food hygiene. Other entries may be found in the main text. The European Single Market in 1993 will have an effect on British food legislation. Many regulations will be altered or will be obsolete.

Act of Parliament Statute law made by Parliament setting out broad principles of law and often empowering the appropriate Minister to make regulations dealing with the law in more detail. *See also*: Food Act 1984, Food hygiene regulations (various), Food safety Act 1990.

Food Safety Act 1990 A major piece of consumer protection legislation which modified and repealed existing law. The main provisions include:

1) **Definitions** (Sec 1)
   a) Food
      - Anything consumed for food including drinks, chewing gum, water and all ingredients used including additives such as flavourings and colourings.
      - Live animals, except when eaten live such as oysters, are excluded.

   b) Food businesses – do not have to make profit, so non-commercial premises such as hospitals and charities are included.

   c) Food sources – are defined as growing crops or live animals. There is a power to make regulations governing these at a later date.

   d) Contact materials – packaging materials and all food equipment may be subject to later regulations.

2) **Sale** (Secs 2, 8, 9, 14, 15)
   Food given away or offered as a prize, food advertised for sale, possessing food for sale, as well as actually selling food may be included.

3) **Human consumption** (Sec 3)
   - All food kept or sold which is normally for human consumption is deemed to be for that purpose unless proved otherwise.
   - Food in a proper waste bin, for example, would not be for human consumption even though it was on food premises.

4) **Injury to health** (Sec 7)
   - This includes temporary and permanent injury as well as cumulative effects caused to a person by eating food.
   - Making food injurious may be caused by adding substances to food, removing them from food or by a process which affects the food. Offences under this heading may result in fines of up to £20 000 and/or up to two years in prison.

5) **Selling food not complying with food safety requirements** (Sec 8)
   - Food which is sold, intended for sale, put on sale or advertised for sale and which does not comply with any food safety requirement causes an offence to be committed.
   - For example, food which is injurious to health, unfit for human comsumption or contaminated would not comply, neither would animals killed in a knacker's yard.

6) **Parts of batches** (Sec 8)
   The whole of a batch, lot or consignment may be judged on the condition of one part.

7) **Power to inspect and seize food** (Sec 9)
   - The local authority has a legal right to perform routine inspections or to act on complaints from the public.
   - These may occur at any reasonable time and at any place including vehicles.
   - The food, if judged not to comply with food safety requirements, may be seized or detained by serving a legal notice.
   - The food may then not be moved or used and may be taken to be destroyed.
   - There is a right of appeal with compensation payable if the actions are found to be unjust.

8) **Improvement notice** (Sec 10)
   - If conditions are found which do not comply with food safety requirements and place food at risk, then an improvement notice may be issued by the local authority to the person running a food business.
   - It states what is wrong, why it is wrong and what must be done to put matters right.
   - Times must be specified for work to be done (minimum 14 days) and for appeals to be lodged.
   - It is an offence not to comply with an improvement notice.

9) **Prohibition order** (Sec 11)
- After the proprietor of a food business has been convicted of a food safety offence, an order prohibiting the premises, parts of the premises, individual processes or persons from involvement in a food business may be issued.
- The notice is conspicuously fixed to the premises and remains in force until the local authority issues a certificate.

10) **Emergency prohibition notices and orders** (Secs 12, 13)
- Without going to court, an authorised officer such as an EHO may issue a notice having immediate effect, and have this confirmed by applying to a court within three days for a court order.
- Such a notice could be used to close a premises if there is imminent risk to health.
- In other respects this is similar to a prohibition order.

11) **Emergency control order** (Sec 13)
In national or widespread food safety emergencies, the Minister may make such an order for example to stop a process or to have food seized or destroyed on a large scale.

12) **Food not of the nature, substance or quality** (Sec 14)
- An offence is committed when food which is not what the general public would normally expect or demand, is sold.
- The legal term 'prejudice' is often used, meaning that the purchaser has been sold inferior goods.
- The maximum fine under this section is £20 000.

See entries in main dictionary for examples of nature, substance and quality.

13) **False descriptions** (Sec 15)
- Food labels, descriptions, menus and advertisements must accurately describe foods.
- The maximum fine under this section is £2 000.

14) **Power to make regulations** (Secs 16–19, 26, 48, 53 and 59)
Regulations covering the following are all possible under the Act:
- composition of food
- microbiological standards
- processes
- practices
- premises
- labelling

- food-contact materials
- 'novel' foods
- special milk designations
- registration and licensing of food premises.

15) **Defences** (**Secs 20–22**)
- Possible defences include proving that the offence was due to the act or default of another person or that all due diligence was exercised.
- Due diligence implies that all reasonable precautions were taken.

16) **Other provisions** (**Sec 23 onwards**)

a) Local authorities may provide training courses in food hygiene.
b) Public analysts must be appointed.
c) Authorised officers may take samples, photographs and may seize food.
d) Food sampling methods are specified.
e) Entry to food premises must be permitted at all reasonable times, obstruction of authorised officers is an offence.
f) Appeals against notices and prosecutions are detailed as are the ways in which legal action is to proceed.
g) Crown Immunity is partially removed.

Food Hygiene (General) Regulations 1970. (Amended 1990/91)

These regulations were originally made under the Food and Drugs Act 1956 and deal with hygiene in all food premises with the exceptions of dairies, slaughterhouses, markets, stalls and delivery vehicles.

The main purpose of the regulations is to protect food from contamination.

1) **Regs 7 and 9**

All equipment, containers and wrappings for food use must not expose food to risk of contamination.

2) **Reg 9**

a) Unfit food must be kept apart from food for human consumption.
b) Food outside must be at least 18 inches off the ground.
c) Open food must be protected from contamination during delivery or sale.

3) **Reg 10**
Food handlers must:
   a) wear clean and washable overclothing
   b) cover cuts and abrasions with a suitable waterproof dressing
   c) not smoke or spit in food rooms.

4) **Reg 13**
Food handlers with certain diseases such as typhoid fever or food poisoning infections must notify their supervisor who should notify the local authority.

5) **Reg 15**
Sanitary conveniences must be:
   a) clean, well lit and ventilated
   b) have a notice clearly displayed requesting food handlers to wash their hands.

6) **Regs 17, 18, 21**
   a) A constant supply of clean water is required at food premises.
   b) Wash hand basins with soap, nailbrush, hot water (cold if no open food is handled) and suitable drying facilities must be provided and only used for hand washing.
   c) Separate sinks with hot and cold water are required for washing food and equipment.

7) **Reg 19**
An adequate supply of first aid materials must be available.

8) **Reg 20**
Suitable accommodation for outer clothing must be provided. Outer clothing must not be kept in a food room unless placed in a locker or cupboard etc.

9) **Regs 22, 23**
Food rooms must be adequately lit and ventilated.

10) **Reg 25**
Food rooms should be constructed so that pests are denied access.

11) **Reg 26**
Waste and refuse should not be allowed to accumulate in food rooms. Adequate storage for refuse and waste must be provided.

Food Hygiene (Amendment) Regs 1990/91 These state the temperature control requirements for high-risk foods.

I

Foods in the following list (from 1 April 1991) must not be kept above 84 °C or below 63 °C:

a) soft cheeses ripened with moulds or micro-organisms
b) cooked foods containing meat, fish, eggs (or substitutes for any of these), cheese, cereals, pulses and vegetables
c) smoked or cured fish
d) smoked or cured meat which has been cut or sliced after smoking or curing
e) desserts containing milk or milk substitutes with a pH of 4.5 or more
f) prepared vegetable salads
g) cooked pies, pasties and sausage rolls or similar products containing meat, fish (or substitutes) or vegetables, unless consumed on the day of production or the day after
h) uncooked or partly cooked pastry or dough products containing meat, fish or their substitutes
i) sandwiches and similar products containing foods in **a**) and **b**) above
j) cream cakes.

There is time allowed for heating food (no unavoidable delay) and also time for service (this may be up to four hours, see **III** below) during which temperature control may lapse.

In the case of equipment breakdown, unavoidable steps in a production process or defrost cycles, a rise of 2 °C for up to two hours is allowed.

II

From 1 April 1993 certain foods must not be kept above 5 °C. These include:

a) cut soft cheeses ripened with moulds or micro-organisms
b) cooked hard or soft cheese in a product
c) cooked meat, fish, eggs (or substitutes for any of these), cereals, pulses and vegetables intended for consumption without reheating
d) smoked or cured fish, whole or cut
e) smoked or cured meats when sliced or cut
f) prepared salads containing any food in **c**) above
g) sandwiches and similar products containing food as in **a**) to **d**) above not intended for sale within 24 hours.

(Note: Sandwiches intended for sale within four hours are excluded from all temperature control while those intended for sale within 4 to 24 hours may be kept at 8 °C.)

III

Foods exempt from temperature control:

a) bread, cakes, biscuits and pastry (including those containing milk or egg added before baking)
b) ice-cream subject to the Ice-Cream (Heat Treatment etc.) Regs 1959
c) canned, dehydrated, pickled or preserved foods
d) uncooked bacon or ham
e) dry pasta or dried mixes
f) chocolate and sugar confectionery
g) milk which is not combined with other ingredients
h) any food intended for sale within two hours after preparation at 63 °C or above
i) any food intended for sale within four hours after preparation at less than 63 °C
j) food displayed for sale on catering premises, in reasonably necessary quantities, for up to four hours for consumption on the premises.

IV

Delivery vehicles (food in)

a) From 1 April 1991, vehicles of more than 7.5 tonnes gross weight must maintain temperatures at no more than 8 °C or no less than 63 °C.
b) From 1 April 1992, vehicles of less than 7.5 tonnes gross weight – as for **a**).
c) From 1 April 1993:
   i) vehicles of more than 7.5 tonnes gross weight must maintain temperatures at no more than 5 or 8 °C (depending on food carried, see above) or no less than 63 °C
   ii) vehicles of less than 7.5 tonnes, if delivery takes more than 12 hours as **i**) above
   iii) if delivery takes less then 12 hours then as for **a**) above.

Food Labelling Regulations 1984, as amended 1990 Most food for human consumption (see exemptions below) must be labelled with:

- name of the food
- ingredients list
- shelf-life ('best-before' date) or in the case of high-risk or perishable food a 'use-by' date
- any special storage information
- name and address of supplier or packer.

Exempted foods include:

- uncut fresh fruit and vegetables
- alcoholic drinks
- bread
- flour confectionery
- vinegar
- sugar
- chewing gum.

Frozen foods, ripening cheese and foods with a shelf-life exceeding 18 months are also exempt until 20 June 1992.

Milk and dairies legislation

- A large number of regulations are in existence dealing with registration, licensing and hygiene standards in milk, milk products, ice-cream and egg producers.
- The heat treatment and labelling of milk is closely specified.
- The sale of raw milk is now restricted to farm visitors or to the ultimate consumer by way of a licensed distributor.
- It cannot be sold for resale to a restaurant for example.
- Since 1990 it has become an offence to sell ungraded eggs with cracks in their shells.

Slaughterhouse and meat trades law

- General hygiene standards are similar to other food premises with EC standards being very strict.
- All meat for human consumption must have been killed in a licensed slaughterhouse and inspected in the prescribed way by authorised officers appointed by the local authority.
- Unfit meat is stained with a black dye at the slaughterhouse and placed in special labelled containers for removal.

Control of infectious diseases

- Food poisoning must be notified to the local authority by medical practitioners or by persons in control of food businesses.
- A food handler or any other person may be excluded from work if they have certain infectious diseases, such as typhoid fever or food poisoning.

Public Health (Control of Diseases) Act 1984 and Public Health (Infectious Diseases) Regs 1963 The Medical Officer of Environmental Health (CCDC) may order a person not to handle food until an infectious disease they are suffering from has been cured. The person may claim for loss of earnings.

Health and Safety at Work, etc Act 1974

- Legislation requiring employers to ensure safe conditions in workplaces for employees and others and to prevent danger to the public caused by workplace activities.
- There are a huge number of regulations made under this and other safety laws, including reporting of accidents and injuries, COSHH, electricity, radiation, lead and noise.

Food Premises (Registration) Regs 1991 *See*: Registration (main text).

Scottish Legal System There are many historical differences between the Scottish and English legal systems at all levels. Most alleged food hygiene offences in Scotland are heard in the Sheriff Court, the Sheriff being a full-time, legally qualified official. Each Sheriff Court has cases referred by an administrative department headed by a Procurator Fiscal. More serious matters may be referred to the High Court which has a presiding Law Lord and jury. Fines are similar to those in England with a £2 000 limit in the Sheriff Court but an unlimited maximum fine and/or one year in jail for other courts.

Scottish Food Hygiene Law

1) The Food Safety Act 1990 replaces the Food and Drugs (Scotland) Act 1956.

2) The main Scottish food hygiene regulations were made in 1959 and have been amended several times and are generally similar to the English Regulations. However, differences occur such as those listed below.
   a) Food handlers must wash their hands after using a sanitary convenience.
   b) Fingernails and hands must be kept clean.
   c) There are detailed requirements for washing utensils including air-drying.
   d) There are detailed requirements for washing glasses and equipment.
   e) There are detailed requirements for the reheating of food to above 82 °C.
   f) Various regulations concerning the treatment of gelatin and fillings for baked goods.

3) Licensed premises must have a hygiene certificate issued by the Environmental Health Department.

4) Market stalls, late night caterers, places of public entertainment and street traders must also have a hygiene certificate similar to the above.

5) The sale of raw milk is prohibited in Scotland.

6) There are specific regulations covering milk and dairies, ice-cream and some aspects of egg production.

7) Slaughterhouse hygiene regulations and meat and poultry inspection regulations are similar to those in England but there are additional regulations governing meat distribution and export.

Appendix 2: Scientific Names

Acarina	mites
Acarus siro	flour mite
Agaricus bisporus/brunescens	common white cultivated mushroom
Alexandrium tamarense	alga (poisonous)
Alphitobius laevigatus	black fungus beetle
Amanita phalloides	Death cap mushroom
Amanita verna/virosa	Destroying angel mushroom
Anagasta kuhniella	mediterranean flour moth
Anisakis simplex	roundworm (fish, humans)
Anthrenus species	carpet, furniture and museum beetles
Ascaris lumbricoides	roundworm (pig, humans)
Blatta orientalis	Oriental cockroach (also known as Black bat beetle or Black clock)
Blattella germanica	German cockroach or steam fly
Bruchus ervi	lentil beetle
Bruchus pisorum	dried pea beetle
Buccinium undatum	whelk or buckie
Calandra granarius	grain weevil
Calandra oryzae	rice weevil
Calliphora erythrocephala	blowfly/bluebottle
Cancer pagurus	crab
Cerastoderma edule	cockle
Cheyletus eruditus	mite (parasite of flour mite)
Chlamys opercularis	queen escallop
Chloromyscium thyrsites	protozoan (milky hake)
Chloromyscium pisiformis	protozoan (rabbits)
Chloromyscium serialis	protozoan (rabbits)
Coxiella burnettii	rickettsia (Q fever)
Crangon crangen	common/brown shrimp
Crassostrea angulata	Portuguese oyster
Crassostrea gigas	Pacific oyster
Cryptolestes ferrugineus	Rust-red grain beetle
Cryptosporidium parvum	coccidial protozoan
Ctenocephalides canis	dog flea
Ctenophalides felis	cat flea
Cysticercus bovis	cyst of beef tapeworm
Cysticercus cellulosae	cyst of pork tapeworm
Cysticercus tenuicollis	cyst of dog tapeworm
Dermestes lardarius	bacon/larder beetle
Dibothriocephalus latus	tapeworm (fish)
Dinophysis	marine alga (poisonous)
Diphyllobothrium latum	tapeworm (freshwater fish)
Distoma hepaticum	liver fluke

Distoma lanceolatum	liver fluke
Drosophila melanogaster	fruit fly
Echinococcus granulosus/ veterinorium	tapeworm (dog)
Entamoeba histolytica	protozoan (amoebic dysentery)
Ephestia elutella	warehouse/cocoa/mill moth
Ephestia kuhniella	mediterranean flour moth
Fannia cannicularis	lesser housefly
Fasciola hepatica	liver fluke
Gambierdiscus toxicus	marine alga
Giardia lamblia	protozoan (diarrhoea)
Gnathocerus cornutus	Broad-horned flour beetle
Gnathocerus maxillosus	Slender-horned flour beetle
Gonyaulax catanella	alga (poisonous)
Gonyaulax tamarensis	alga (poisonous)
Hoffmannophila pseudospretella	brown house moth
Homarus gammarus	lobster
Iridomyrmex humilis	Argentine ant
Lasioderma serricorne	Cigarette beetle
Lasius niger	black or garden ant
Lepisma saccharina	silverfish
Liposelcis granicola	booklouse
Littorina littorea	periwinkle
Lucilia sericata	greenbottle
Maestomys ratalensis	multimate rat
Monomorium pharaonsis	Pharaoh's ant
Mus domesticus	house mouse
Musca autumnalis	autumn fly
Musca domestica	housefly
Mytilus edulis	mussel
Nephrops norvegicus	Dublin Bay prawn/langoustine/ Norway lobster/scampi
Neptunea antigua	red whelk
Niptus hololeucus	Golden spider beetle
Nitzschia pungens	marine diatom (poisonous)
Oesophagostum radiatum	roundworm (cow, sheep)
Oncocerca gibsoni	roundworm (cow)
Oryzaephilus surinamensis	Saw-toothed grain beetle
Ostrea edulis	native oyster
Palinurus elephas	crawfish/langoustine/pinkshrimp/ spiny lobster
Pandalus borealis	prawn
Pandalus montagui	pink shrimp
Pecten maximus	escallop
Periplaneta americana	American cockroach

Periplaneta australasiae	Australian cockroach
Phaseolus vulgaris	red kidney bean
Phocanema spp	roundworm (cod)
Phytophthora infestans	downy mildew
Plodia interpunctella	Indian-meal moth
Prorocentrum concavum	marine alga (poisonous)
Prorocentrum mexicana	marine alga (poisonous)
Ptinus tectus	Australian spider beetle
Ptychodiscus brevis	alga (poisonous)
Pulex irritans	human flea
Pyrodinium bahamense	alga (poisonous)
Rattus norvegicus	brown rat
Rattus rattus	black rat
Saccharomyces	yeast
Sarcophoga carnaria	flesh fly
Schistosoma japonicum	blood fluke
Schistosoma mansoni	blood fluke
Sitophilus granarius	grain beetle or weevil
Sitophilus oryzae	rice weevil
Sitophilus zea-mais	maize weevil
Stegobium paniceum	biscuit or drug-store beetle
Taenia hydatigena	tapeworm (dog)
Taenia saginata	tapeworm (cow)
Taenia solium	tapeworm (pig)
Tenebrio molitor	yellow mealworm (beetle)
Tenebrio obscuris	dark mealworm (beetle)
Tenebroides mauritanicus	cadelle beetle
Thaumatomia notata	yellow swarming fly
Thermobia domestica	firebrat
Tineola bisselliella	clothes moth
Toxocara canis	tapeworm (dog)
Toxocara felis	tapeworm (cat)
Tribolium castaneum	Rust-red flour beetle
Tribolium confusum	Confused flour beetle
Tribolium destructor	Dark flour beetle
Trichinella spiralis	roundworm (humans, rat, pig)
Tyrophagus casei	cheese mite
Tyroglyphus farinae	flour mite
Venus mercenaria	clam
Vespula vulgaris	wasp
Xenopsylla cheopsis	rat flea

Note: Names of bacteria appear in the main body of the text.

Appendix 3: Abbreviations

AFD	Accelerated Freeze Drying
ASP	Amnesic Shellfish Poisoning
ATA	Alimentary Toxic Aleukia (Aleukia)
ATP	Adenosine Triphosphate
a_w	Available water
BFMIRA	British Food Manufacturing Industries Research Association
BHA	Butylated Hydroxy Anisole
BHC	Benzene Hexachloride
BHT	Butylated Hydroxy Toluene
BPCA	British Pest Control Association
BS	British Standard
BSI	British Standards Institution
BSE	Bovine Spongiform Encephalopathy
CAMPDEN	Campden Food Preservation Research Association
CDR	Communicable Disease Report
CDSC	Communicable Disease Surveillance Centre
CIP	Cleaning-In-Place
COSHH	Control Of Substances Hazardous to Health Regs 1988
CPU	Central Processing Unit
CPHL	Central Public Health Laboratory
DAFS	Department of Agriculture & Fisheries for Scotland
DDT	Dichlorodiphenyl Trichloroethane
DHSS	Department of Health and Social Security
DoH	Department of Health
DSP	Diarrhetic Shellfish Poisoning
EC	European Community
EHO	Environmental Health Officer
FDF	Food and Drink Federation
FMBRA	Floor Milling and Baking Research Assocaition
FOOD RA	Leatherhead Food Research Association
HACCP	Hazard Analysis Critical Control Points
HASAWA	Health and Safety at Work, etc Act 1974
HAZOP	Hazard Operability Studies
HAV	Hepatitis type A Virus
HBV	Hepatitis type B Virus
HCH	Hexachlorocyclohexane
HSC	Health and Safety Commission
HSE	Health and Safety Executive
HTST	High Temperature Short Time
IEHO	Institution of Environmental Health Officers
IFST	Institute of Food Science and Technology
ITSA	Institute of Trading Standards Administration
ISO	International Standards Office

IVS	Intervening Ventilated Space
LACOTS	Local Authorities Co-ordinating Body on Trading Standards
LC_{50}	Lethal Concentration (50%)
LD_{50}	Lethal Dose (50%)
MAFF	Ministry of Agriculture, Fisheries and Food
NCC	National Consumer Council
NFU	National Farmers' Union
NHS	National Health Service
NSP	Neurotoxic Shellfish Poisoning
PDA	Potato Dextrose Agar
pH	Hydrogen ion index (measure of acidity/alkalinity)
PHLS	Public Health Laboratory Service
QA	Quality Assurance
QAC/QUATS	Quaternary Ammonium Compound
QC	Quality Control
PSP	Paralytic Shellfish Poisoning
REHIS	Royal Environmental Health Institution of Scotland
RH	Relative Humidity
RIDDOR	Reporting of Injuries, Diseases and Dangerous Occurrences Regs 1985
RIPHH	Royal Institute of Public Health and Hygiene
RSH	Royal Society of Health
TB	Tuberculosis
TMA	Trimethylamine
SOFHT	Society of Food Hygiene Technology
UHT	Ultra Heat Treated
USFDA	United States Food and Drug Administration
UV	Ultra Violet (light)
WC	Water Closet (lavatory)
WHO	World Health Organisation
WO	Welsh Office

Appendix 4: Major causes of food poisoning and food-borne disease

Bacterium	*Onset period*	*Duration*	*Symptoms*	*Foods commonly involved*	*Comments*
Salmonella species Gram-negative non-sporing bacillus; facultative anaerobe	12 to 36 hours (unusually 6 to 72 hours or more) [Infection]	1 to 7 days	Diarrhoea, vomiting, fever, headache and aching of limbs	Meat, poultry and their products, eggs, raw milk, milk products	Widespread; found in all farm and wild animals, pets, human and animal excreta
Staphylococcus aureus Gram-positive non-sporing coccus; facultative anaerobe, growing best in presence of oxygen	1 to 6 hours [Exotoxin in food]	6 to 24 hours	Severe vomiting, diarrhoea, abdominal pain/cramps, sometimes collapse	Handled dishes, e.g. cold meats, trifles, cream-filled cakes	Contamination most often from food handlers; most salt- and sugar-resistant organism; refrigeration of susceptible foods essential
Clostridium perfringens Gram-positive sporing bacillus; anaerobe	8 to 22 hours [Endotoxin in intestine]	12 to 48 hours	Mild, abdominal pain and diarrhoea; vomiting rare but nausea common	Meat dishes, e.g. stews and rolled joints	Survives cooking, hence rapid cooling and refrigeration essential if food not for immediate consumption; thorough re-heating essential; contamination usually from excreta, raw meat, poultry, soil
Bacillus cereus Gram-positive sporing bacillus; aerobe	1 to 16 hours [Commonly exotoxin in food]	12 to 48 hours	Similar to *Staphylococcus aureus*	Cooked rice, especially boiled and eaten cold, re-heated, fried	Widely distributed, found in soil, cereals (especially rice), vegetables, dairy products; refrigeration of cooked rice essential
Campylobacter jejuni/coli Gram-negative non-sporing spirally-curved bacillus; microaerophilic to anaerobic	2 to 11 days [Enterotoxin in intestine]	1 day to several weeks	Start as influenza-like illness with diarrhoea (usually containing blood), abdominal pain, nausea	Mainly raw poultry but also meat, offal, raw milk, contaminated water	Contamination usually from human and animal excreta (especially birds); complications include meningitis and endocarditis

Appendix 4: Major causes of food poisoning and food-borne disease – continued

Bacterium	*Onset period*	*Duration*	*Symptoms*	*Foods commonly involved*	*Comments*
Listeria monocytogenes Gram-positive non-sporing bacillus; facultative anaerobe	Up to 3 months [Infection]	Varies; often long-lasting	Fever, with many possible complications, including meningitis, septicaemia, abscesses; pregnant women at risk, with miscarriage and still-births possible	Commercial chilled foods kept at above 5°C, e.g. prepared salads, sandwiches, cooked meats and ready-to-eat meals	Widespread; found in humans, all types of animals, sewage, soil; capable of slow growth at temperatures as low as 2 to 3°C
Escherichia coli Gram-negative non-sporing bacillus; aerobe	12 to 72 hours [Infection and enterotoxin]	1 to 7 days	Watery diarrhoea, vomiting abdominal pain, sometimes fever	Raw foods of animal origin, e.g. meat, poultry, fish and their products	Common cause of traveller's diarrhoea; infants and young children at risk; spread through poor hygiene practices: sewage and water treatments effective controls
Vibrio parahaemolyticus Gram-negative non-sporing curved bacillus; facultative anaerobe	Usually 12 to 18 hours [Infection]	2 to 5 days	Burning sensation in stomach, abdominal pain, vomiting, watery diarrhoea	Shellfish	Usually inhabits coastal salt waters; one of the commonest causes of food poisoning in Japan
Shigella species Gram-negative non-sporing bacillus; facultative anaerobe	1 to 2 days [Infection and toxins]	Few days	Diarrhoea, vomiting, abdominal pain and cramps, occasionally (if severe) blood and pus in faeces	Any food contaminated with untreated sewage; contaminated water	Intestinal organism (humans and animals); control of flies, good personal hygiene, treatment of water and sewage reduce spread
Clostridium botulinum Gram-positive sporing bacillus; anaerobe	18 to 36 hours [Exotoxin – a neurotoxin – in food]	Death in 24 hours to 8 days or slow convalescence over several months	Slurred speech, double vision, loss of balance; death follows interference with control of breathing	Any canned or bottled food of pH above 4.5, and smoked foods, especially fish	Widely found in nature, including soil, meat and fish; the most heat-resistant of all food poisoning organisms; outbreaks rare in UK, not uncommon in USA